AF843944

An Introductory Guide to Quantitative Data Analysis and Interpretation using SPSS 30

An Introductory Guide to Quantitative Data Analysis and Interpretation using SPSS 30 is a comprehensive guide to quantitative analysis. It integrates knowledge of research methods, data analysis, and report writing, including data exploration, data preparation, and analysis.

This textbook provides readers with fundamental knowledge of quantitative data analysis through a sample of research objectives, step-by-step guidance for generating SPSS outputs, hypothesis testing, and interpreting analysis outputs. It focuses on frequently used descriptive, bivariate, and multivariate analyses and discusses the assumptions and requirements for these methods. Using SPSS version 30 and also providing a complete case study, this book will help readers to get a big picture of how to get the collected data ready and complete further analysis. This is followed by a step-by-step guide to interpreting the outputs to match research objectives or hypothesis testing. Lastly, it provides a sample write-up for a research article or dissertation using APA format.

This guide will be an essential companion for beginner or university students who are learning quantitative data analysis using SPSS software and/or doing their research project. It will also be beneficial to professionals and researchers who want to revive their quantitative data analysis and format their academic papers to maintain clarity and professionalism in scientific writing using APA format.

Yeoh Sok Foon is an independent researcher in Malaysia. She previously worked as a lecturer teaching statistics, quantitative methods, and management. She has varied research and publication experience.

An Introductory Guide to Quantitative Data Analysis and Interpretation using SPSS 30

Outputs and Interpretation for APA Format

Yeoh Sok Foon

Routledge
Taylor & Francis Group
LONDON AND NEW YORK

Designed cover image: Getty Images

First published 2026
by Routledge
4 Park Square, Milton Park, Abingdon, Oxon OX14 4RN

and by Routledge
605 Third Avenue, New York, NY 10158

Routledge is an imprint of the Taylor & Francis Group, an informa business

British Library Cataloguing-in-Publication Data
A catalogue record for this book is available from the British Library

ISBN: 978-1-041-14314-7 (hbk)
ISBN: 978-1-041-13776-4 (pbk)
ISBN: 978-1-003-67390-3 (ebk)

DOI: 10.4324/9781003673903

Typeset in Optima
by Apex CoVantage, LLC

Access the Support Material: www.routledge.com/9781041137764

Contents

Tables

Figures

Preface

Data analysis is part of the big puzzle of the research process to assist in reaching your research findings or generating new ideas in your field of interest. What should you do in data analysis? It involves a systematic process of selecting and applying statistical methods to get your data ready to be used in data description and illustration, and evaluation.

Research is a systematic process to inquire into new knowledge or to prove a theory. In fact, we should learn what data analysis is before starting a research project to ensure the success rate and to have implementable research. It will affect the research design, instrumentation, questionnaire design, and data collection methods.

Who needs this book?

This book is beneficial for those who are beginners or want to refresh their knowledge on data analysis. The main idea of this book is to share how to select appropriate data analysis methods, provide detailed screenshots of steps to run an analysis using IBM, SPSS version 30, provide guides to interpret the relevant outputs, and provide a sample write-up in APA (7th ed.) format.

What should you have/know before starting?

It is assumed that you have installed IBM, SPSS on your PC or laptop, done data entry, and have a basic understanding of the scale of measurement.

Let's explore the list of steps and techniques to succeed in the data analysis. Practice makes perfect. I wish you a great experience in learning new knowledge.

1 Exploratory data analysis (EDA)

Learning outcomes

At the end of this section, participants will be able to:

- Recognise the purposes of EDA,
- Demonstrate EDA step by step using SPSS, and
- Interpret the EDA outputs.

1.1 Purpose of performing EDA

The following are the purposes of performing EDA analysis:

- Generate descriptive statistics and also graphics to provide an overall view of a data set.
- Check the possible existence of missing data and outliers.
- Test the assumption of normality in the data set.

Scenario

Mary wants to maximize the view of her data set and also check the data distribution of the study variables before hypothesis testing.

Research Objective: To identify the significant difference between gender and shopping behaviour.

Independent variable: Gender
Dependent variable: Shopping behaviour

** Shopping behaviour data were collected using a total of 15 questions with 5-point Likert scales (strongly agree to strongly disagree). There are two questions with negative statements, and remember that the student needs to recode the item. Then, the summation score was tabulated as a new variable using the computation function before EDA analysis.*

DOI: 10.4324/9781003673903-1

1.2 Steps to perform EDA

1. On the menu bar, select **File**, then **Open**, and click **Data**. See Screen 1.1.

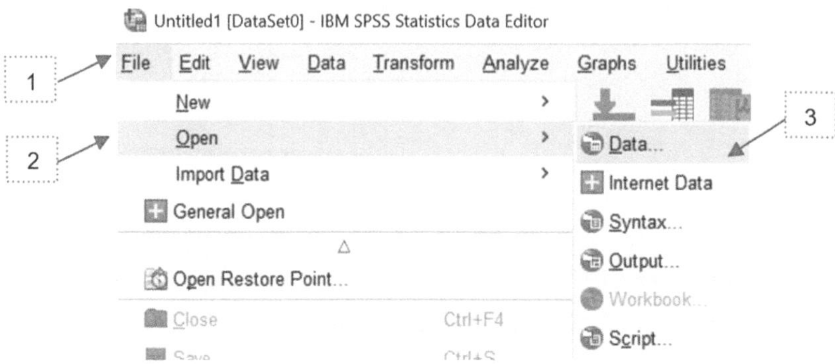

Screen 1.1 File options

2. The Open dialogue box appears (see Screen 1.2). In the sample data folder, select the data file **Shopping behaviour. sav.** and click **Open** to open this data file.

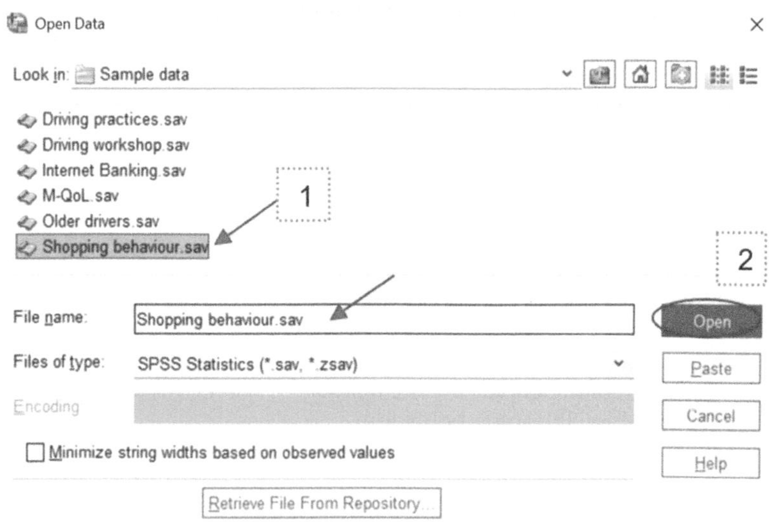

Screen 1.2 Open data

3. Go back to the menu bar, and select **Analyze**, followed by **Descriptive Statistics** (see Screen 1.3).

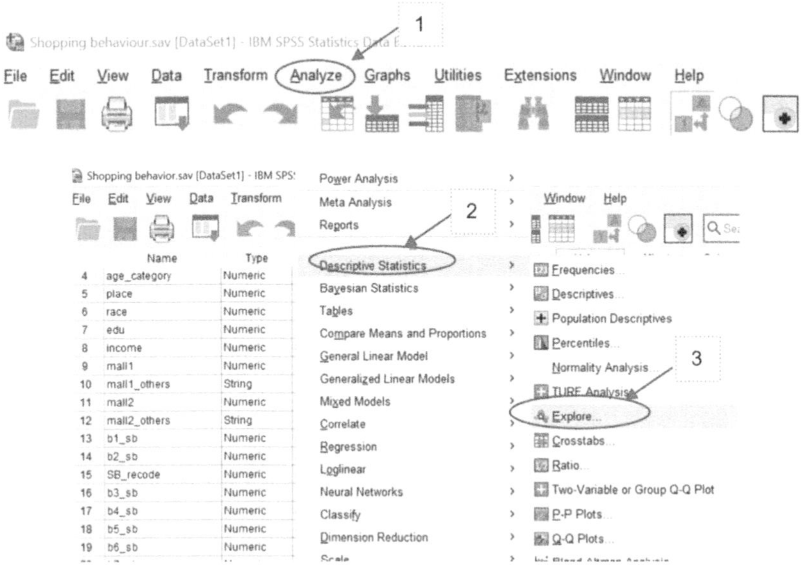

Screen 1.3 Descriptive under the Analyze function

Next, click **Explore**, and Screen 1.4 appears.

Screen 1.4 Explore

4. Select the dependent variable (SB_sumscore) from the variable list and enter it into **Dependent List** (see Screen 1.5).

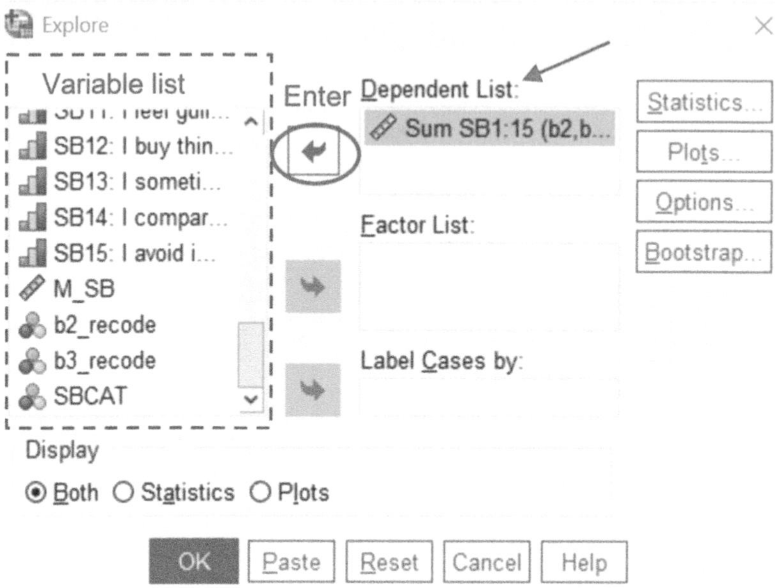

Screen 1.5 Explore: Insert variable to Dependent List

5. Select the factor variable GENDER and insert it into the **Factor List** (see Screen 1.6). **Label Cases by** an identification variable (ID).

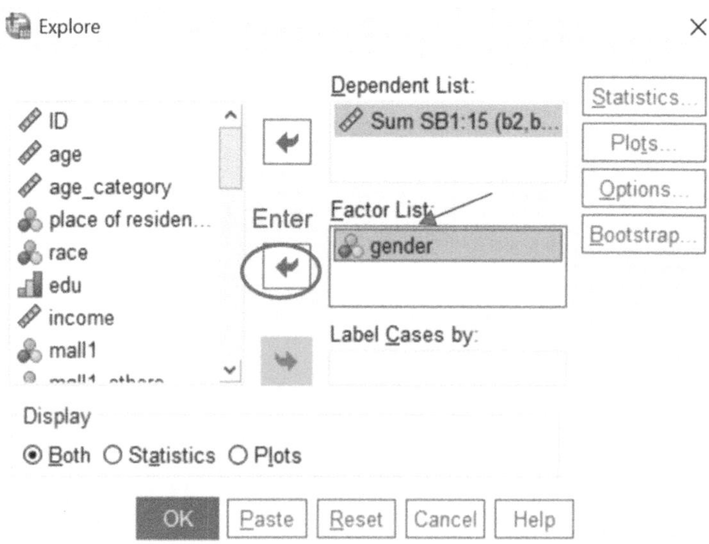

Screen 1.6 Explore: Insert variable to Factor List

6. Click the **Statistics** button, and Screen 1.7, **Explore: Statistics**, opens up. Check ☑ Statistics, including **Descriptives**, **M-estimators**, **Outliers,** and **Percentiles** in the **Explore: Statistics** dialogue box and select **Continue**.

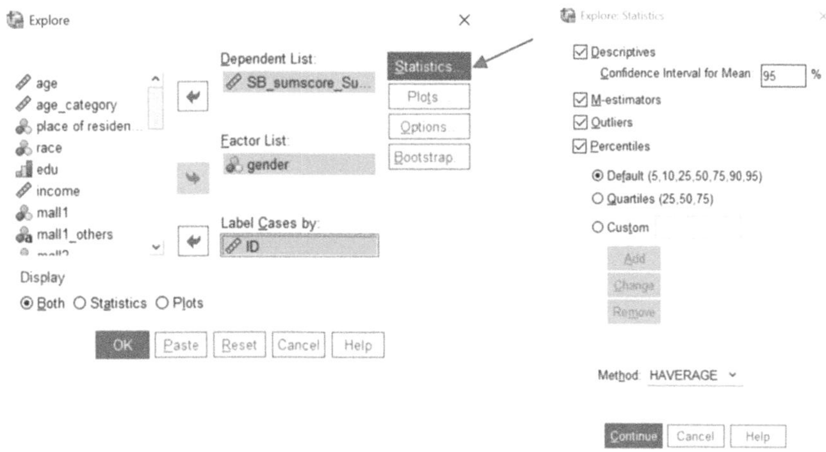

Screen 1.7 Explore: Statistics window

8. Click the **Plots** button on the **Explore** dialogue box. Screen 1.8 opens up.

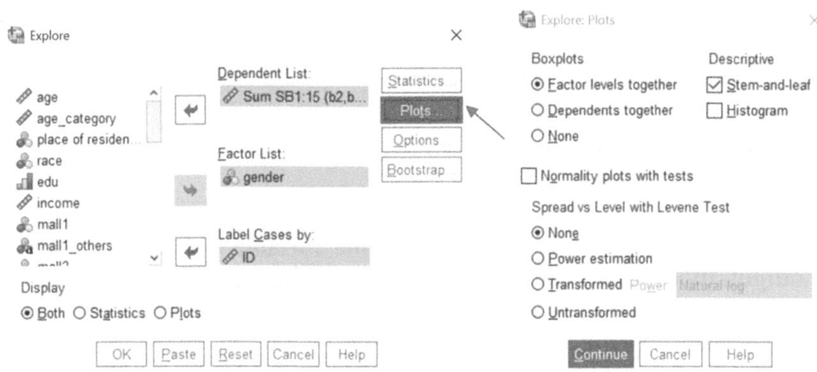

Screen 1.8 Explore: Plots

9. In the **Explore: Plot** dialogue box (see Screen 1.9), under

   i. **Boxplots,** select "Factor levels together."
   ii. **Descriptive**, select "Stem-and-Leaf" and "Histogram."
   iii. Check ☑ **Normality plots with tests**.
   iv. **Spread-versus level with Levene Test**, select **Power estimation**.

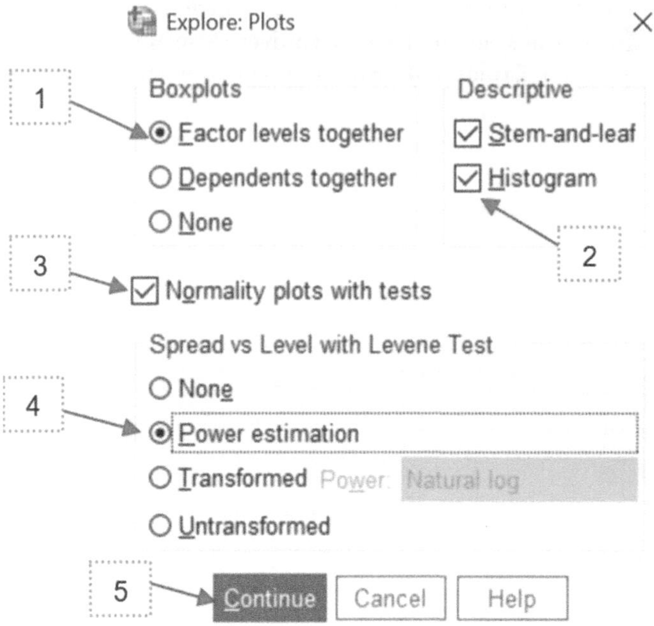

Screen 1.9 Explore: Plots options

10. Lastly, click **Continue**, followed by **OK**, to run EDA.

1.3 Interpretation of the selected SPSS outputs of exploratory data analysis (EDA)

Independent variable: Gender (male and female)
Dependent variable: Shopping behaviour (summation score of 15 items)

In SPSS 30, label is shown instead of the variable name. In the SPSS output, the students can double-click it and change to the desired variable name. The SPSS output can be copy and paste as an editable table. See Screen 1.10 for the sample screenshot.

1. Interpret the 25th and 75th percentiles of shopping behaviour by gender, based on the percentiles output.

   Female shoppers perceived higher (more) in shopping behaviour (*M* = 50.00 to 60.25) compared to male shoppers (*M* = 48.00 to 53.00).

			Percentiles						
		Gender	5	10	25	50	75	90	95
Weighted Average(Definition 1)	Shopping behaviour Sum Score	Male	41.00	44.70	48.00	51.00	53.00	56.30	57.15
		Female	44.95	45.00	50.00	53.00	60.25	64.20	66.05
Tukey's Hinges	Shopping behaviour Sum Score	Male			48.00	51.00	53.00		
		Female			50.00	53.00	60.00		

Screen 1.10 Percentiles

If the value of the 5% trimmed mean varies a lot from the mean, it is a sign that outliers exist.

A histogram is a graphical method to assess the data distribution (see Screens 1.13–1.14). The students can edit the histogram by double-clicking on it; see the sample in Screen 1.12. The chart editor dialogue box appears; the variable name and text size and colour can be edited. The students can also add the normal distribution curve in the chart editor.

2. Refer to the descriptive output (Screen 1.11) and two histograms (Screens 1.13–1.14). Determine whether the arithmetic mean of shopping behaviour is a good descriptor for male and female shoppers.

	Gender			Statistic	Std. Error
Shopping behaviour Sum Score	Male	Mean		50.67	.716
		95% Confidence Interval for Mean	Lower Bound	49.21	
			Upper Bound	52.12	
		5% Trimmed Mean		50.82	
		Median		51.00	
		Variance		18.457	
		Std. Deviation		4.296	
		Minimum		41	
		Maximum		58	
		Range		17	
		Interquartile Range		5	
		Skewness		-.486	.393
		Kurtosis		-.127	.768
	Female	Mean		54.55	1.065
		95% Confidence Interval for Mean	Lower Bound	52.40	
			Upper Bound	56.71	
		5% Trimmed Mean		54.45	
		Median		53.00	
		Variance		43.065	
		Std. Deviation		6.562	
		Minimum		44	
		Maximum		67	
		Range		23	
		Interquartile Range		10	
		Skewness		.348	.383
		Kurtosis		-.891	.750

Screen 1.11 Descriptives

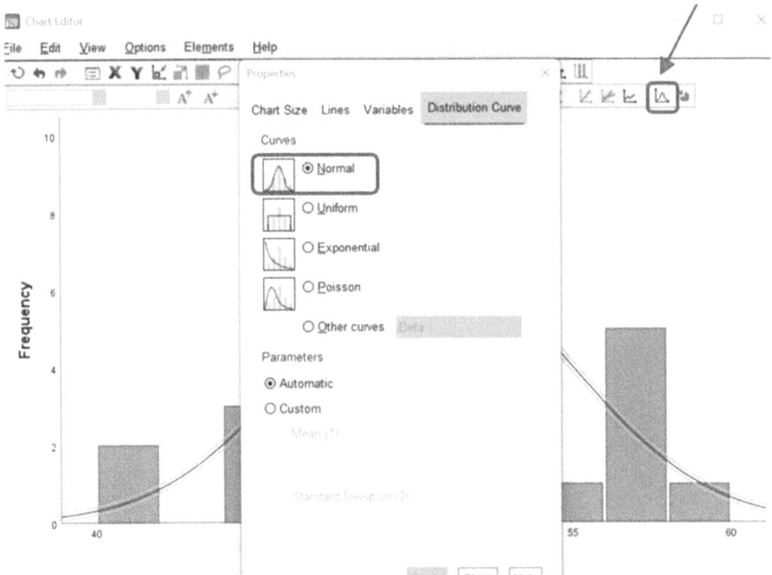

Screen 1.12 Chart editor

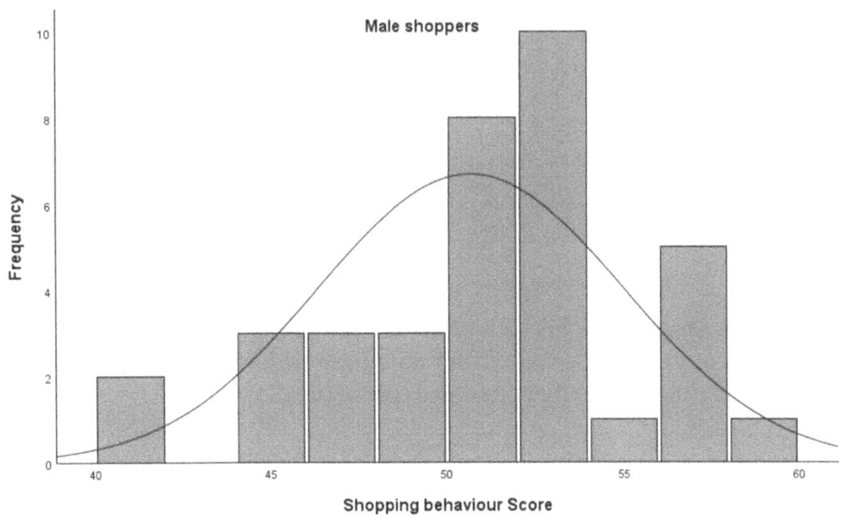

Screen 1.13 Histogram for male shoppers with normal curve

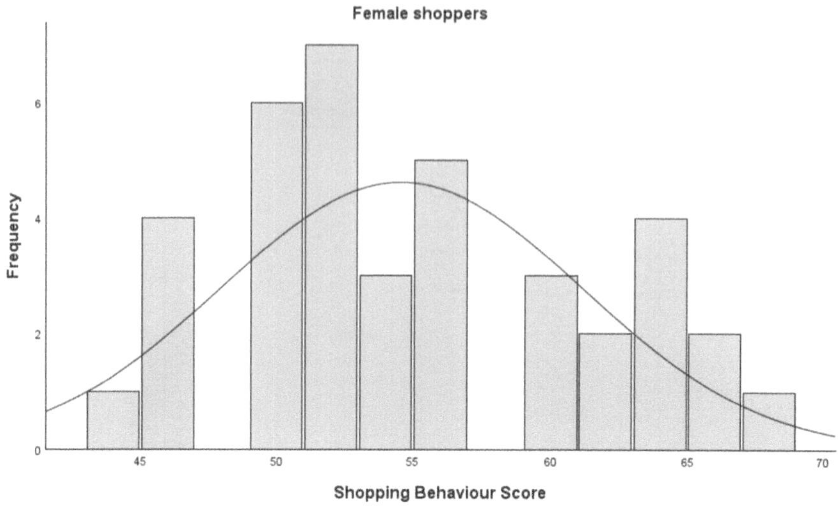

Screen 1.14 Histogram for female shoppers with normal curve

The arithmetic mean is considered a good descriptor of the shopping behaviour for both males and females because the skewness value (*SK* = −0.486 and 0.348) falls within ±2.0. The reference to the cutoff point is set by George and Mallery (2024). The kurtosis is within ±1.0 for male (*K* = −0.127) and female (*K* = −0.891). (Note: If using skewness and kurtosis as the reference for checking normality, this data is considered normal.)

3. Screen 1.15 is the test of normality outputs; it is useful to assess the normality assumption.

> * **Report Shapiro-Wilk if your study sample size is <50;**
> * **Check out the histogram of your data to see if it is normal; boxplot to see the potential outliers and extreme values.**

		Kolmogorov-Smirnov[a]			Shapiro-Wilk		
	Gender	Statistic	df	Sig.	Statistic	df	Sig.
Shopping behaviour Sum Score	Male	.133	36	.110	.960	36	.214
	Female	.127	38	.126	.943	38	.052

a. Lilliefors Significance Correction

Screen 1.15 Test of normality

4. Example of the hypotheses for the tests of normality (symmetrical).
   Null hypothesis

   $H_{o(m)}$: The distribution of mean shopping behaviour for male shoppers is normal.
   $H_{o(f)}$: The distribution of mean shopping behaviour for female shoppers is normal.

   Alternative hypothesis

   $H_{a(m)}$: The distribution of mean shopping behaviour for male shoppers is not normal.
   $H_{a(f)}$: The distribution of mean shopping behaviour for female shoppers is not normal.

5. Decide on the null hypothesis and justify.
   Refer to the Kolmogorov-Smirnov (see Screen 1.15).
   Fail to reject $H_{o(m)}$ because the obtained *p*-value 0.11 ($p > 0.05$) is higher than the alpha value ($\alpha = 0.05$).
   Fail to reject $H_{o(f)}$ because the obtained *p*-value 0.126 ($p > 0.05$) is higher than the alpha value ($\alpha = 0.05$).
6. Conclude on the normality assumption from the test of normality results and normal Q-Q plot outputs.
   Since we support H_o, we can conclude that the distribution of mean shopping behaviour for male shoppers is normal.
   Likewise, $H_{o(f)}$ is supported; the distribution of mean shopping behaviour for female shoppers is normal.
   Hence, both male and female shoppers met the normality assumption.
7. A normal Q-Q plot is another graphical method to assess data distribution (see Screens 1.16–1.17). Describe whether the data is normally distributed.
   The data is normally distributed because the points fall roughly on a straight diagonal line.

Screen 1.16 Normal Q-Q plot for male shoppers

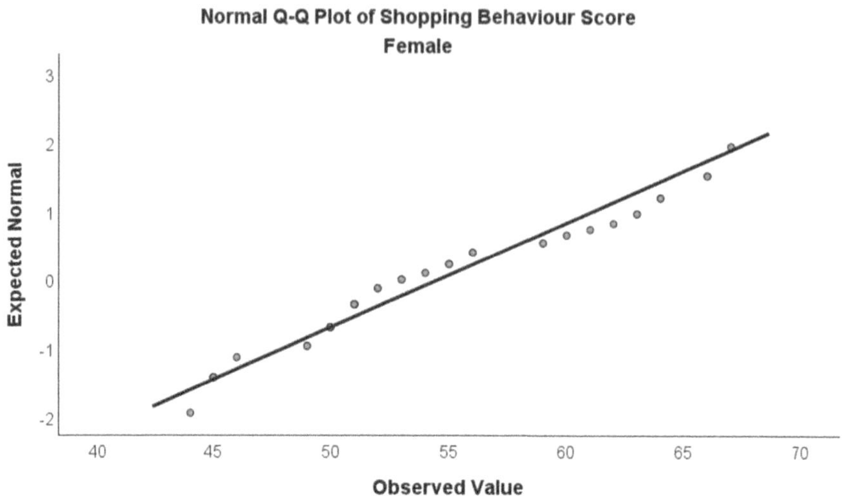

Screen 1.17 Normal Q-Q plot for female shoppers

8. Based on the following boxplot given, compare the shopping behaviour for male and female shoppers.

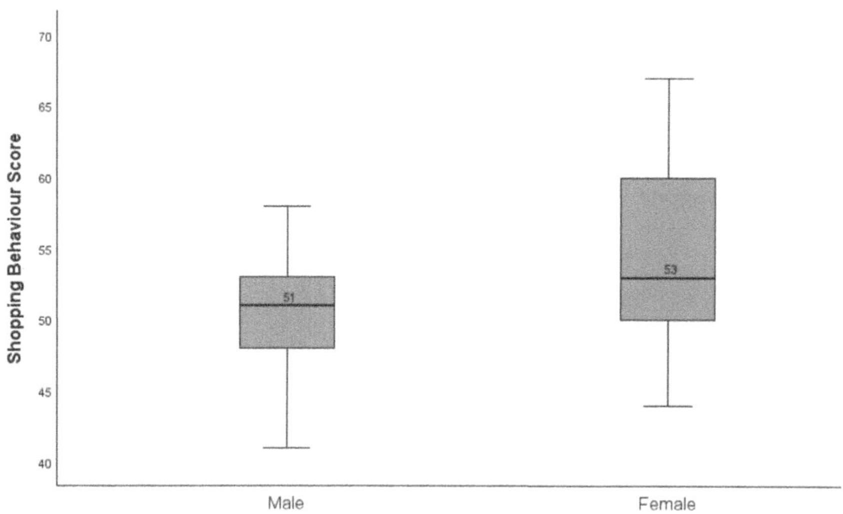

Screen 1.18 Boxplot of shopping behaviour by gender

8a. Compare the median shopping behaviour for male and female shoppers.
In general, the median shopping behaviour for females is slightly higher (*Mdn* = 53.00) than then male shoppers (*Mdn* = 51.00).
8b. Determine the existence of outliers and extreme values.
Check the boxplot, any data that falls outside of the whisker box is an outlier and extreme value with a circle and label asterisk (*), respectively, in case your data set has the presence of outliers and extreme values. Carefully evaluate if there is a need to remove the cases.

A **transformation** is needed when your data is not normally distributed and you expect it to be normally distributed. There are several power alternatives that are available to choose from based on the power estimation result displayed under the spread versus level plot (see Screen 1.19).

Let's say your data is not normally distributed. Re-run the EDA analysis from step 1, and select the transformation option based on the power estimation

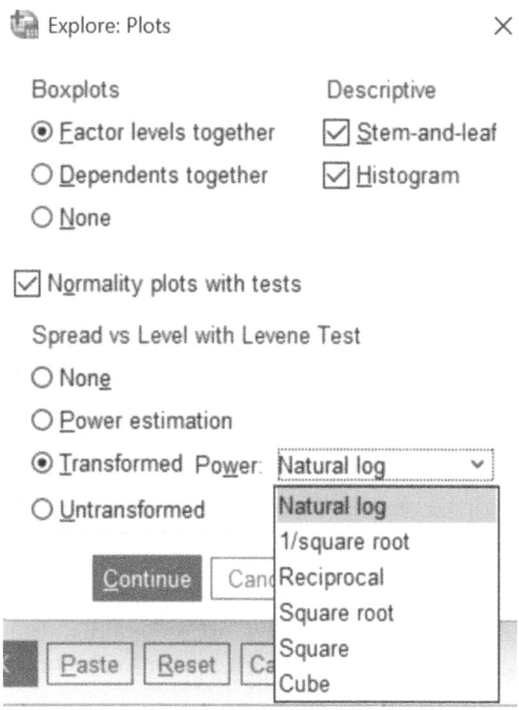

Screen 1.19 Explore: Plots

in step 10 to **transform** your data. Another EDA analysis is needed to check whether the data follow a normal distribution using the new variable generated by the transformation.

See Table 1.1 for the summary of the options for transformations (Osborne, 2002).

Table 1.1 Options for data transformation and their description

Power	*Transformation options*	*Description*
3	Cube	Each data value is cubed.
2	Square	Each data value is squared.
1	No change	No transformation or re-expression is required.
½	Square root	The square root of each data value is calculated.
0	Logarithm/Natural log	Natural log transformation.
−½	Reciprocal of the square root	For each data value, the reciprocal of the square root is calculated.
−1	Reciprocal	The reciprocal of each data value is calculated.

After data transformation for the selected variable, the data may not always turn to normal, but the skewness of the data may be reduced. If the data is normal, the students can choose to run a parametric test to test their hypothesis. Otherwise, the students need to choose non-parametric tests. In general, we expected the data to be normal; thus, this book focuses more on parametric tests.

1.4 Sample write-up of EDA analysis in APA format

This section discusses the data distribution of shopping behaviours and gender. In comparison, female shoppers perceived higher (more) in shopping behaviour (M = 50.00 to 60.25) than male shoppers (M = 48.00 to 53.00). Similarly, the median shopping behaviour for females is slightly higher (Mdn = 53.00) than that of male shoppers (Mdn = 51.00). This implies that female shoppers enjoyed shopping more than male shoppers. The arithmetic mean is considered a good descriptor because the skewness value (SK = −0.486 and 0.348) falls within ±2.0, and the kurtosis is within ±1.0 for male (K = −0.127) and female (K = −0.891) (George & Mallery, 2024). Both male and female shoppers met the normality assumption. Based on the Kolmogorov-Smirnov, both obtained p-values (0.11–0.12) are larger than the alpha value (α = 0.05) and thus fail to reject the null hypothesis. Histogram with normal curve for both genders as presented in Screens 1.13–1.14.

2 Data transformation

Learning outcomes

At the end of this section, participants will be able to:

- Recognise the purposes and various methods of data transformation,
- Perform data transformation using SPSS,
- Know how to replace the missing value, recode a variable, and create a dummy variable,
- Perform data transformation involving mathematical functions,
- Break down a numerical variable into different groups using a class interval, and
- Know the tips to avoid or reduce the mistakes in transforming data.

2.1 Functions of data transformation

This function is necessary to prepare your data for conducting a specific data analysis. These include:

1. Replace the missing value
2. Recode variables (e.g. negative statement, combine groups, etc.)
3. Create a dummy variable (to be discussed at the advanced level)
4. Compute (e.g. calculate a composite score/total scale scores/summation score)
5. Break down a numeric variable into categories/groups

After data entry, we will first carry out an exploratory descriptive analysis to check whether there are missing values and potential outliers and gain an overview of the data set.

DOI: 10.4324/9781003673903-2

2.2 Perform data transformation using SPSS

2.2.1 Replace missing value

In case there is a **missing value** in your existing data set, instead of case-wise deletion, you may choose the function under "Transform" to **replace missing values**. For example, there is a missing value of a monthly income (an open-ended data, ratio scale) in the shopping behaviour sample data.

1. On the menu bar, go to **File**, then **Open**, and then **Data** (see Screen 2.1).

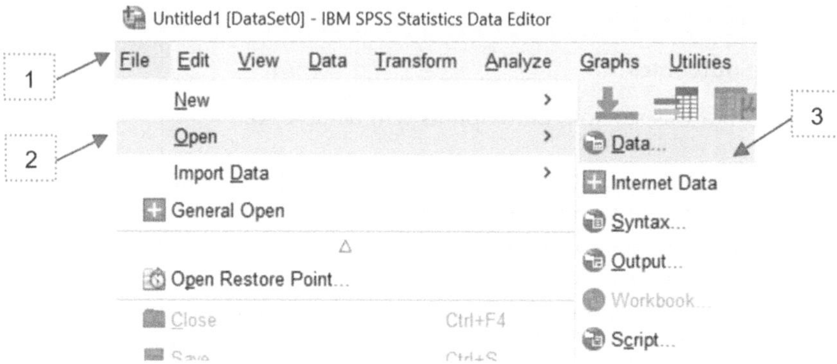

Screen 2.1 File options

2. The **Open Data** dialogue box appears (see Screen 2.2). In the sample data folder, select the **Shopping behavior.sav** and click **Open** to open this data file.

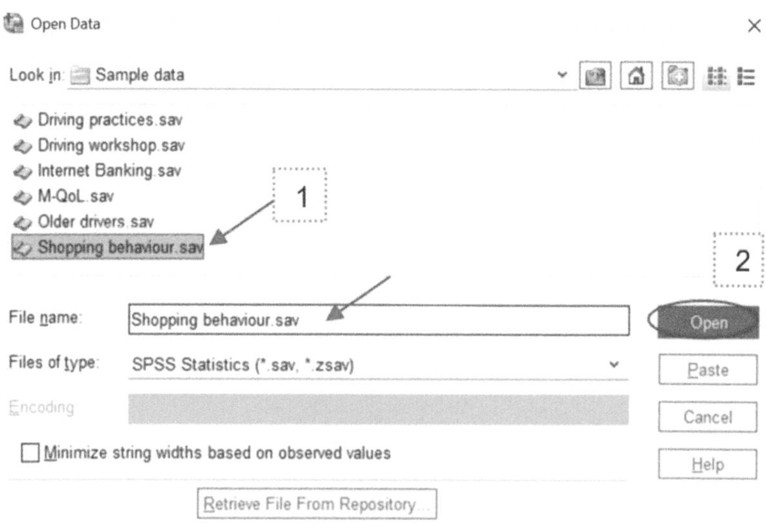

Screen 2.2 Open data

3. Again, go to the menu bar, then select **Transform**, followed by the option **Replace Missing Values . . .** (see Screen 2.3).

Screen 2.3 Transform at the Main Menu Bar

Then, the option of **Replace Missing Values** will appears (see Screen 2.4).

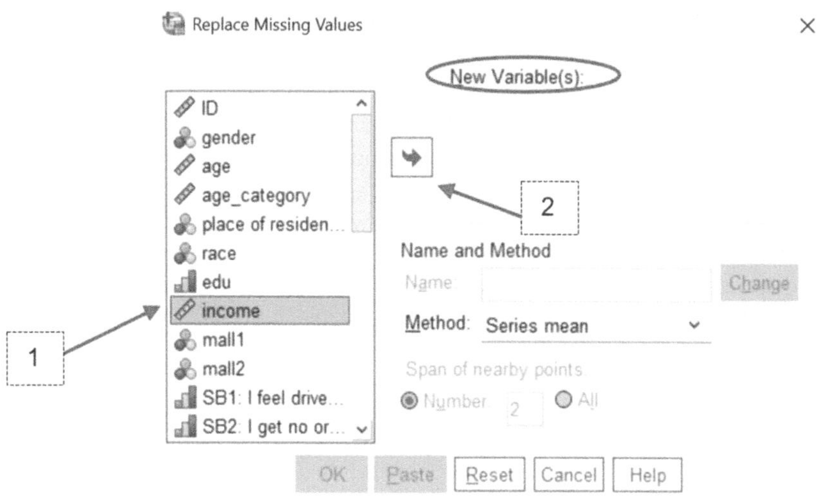

Screen 2.4 Replace Missing Values

You need to select and move the variable that contains the missing value into **New Variable(s)** box. A new variable will be created based on the method you choose to generate the value for missing data. Replacing a value using the method of series mean is default by system. In this example, any missing value of income will be replaced by the mean income, which is the series mean method.

Note that, more than one variable can enter at the same time for the same method. A new variable will be created with a replaced value for the particular respondent based on the chosen method (see Screen 2.5).

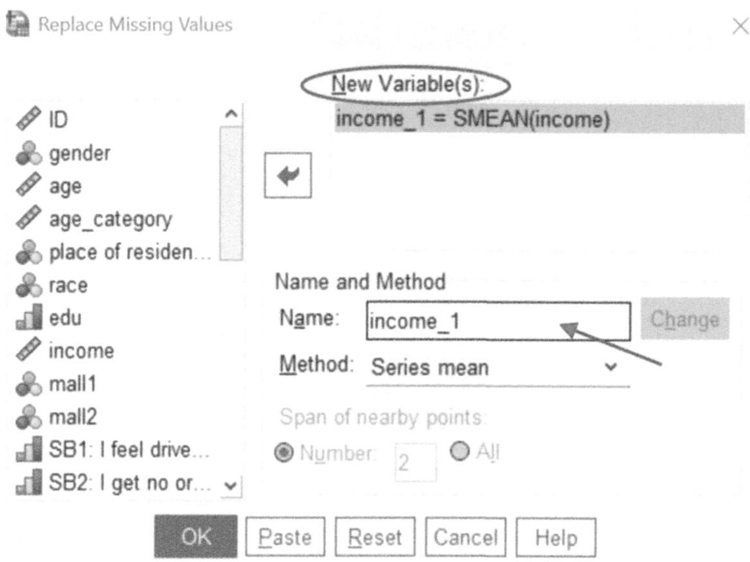

Screen 2.5 Selection method for Replace Missing Values

The new variable name can be changed by double-clicking on the text box. Click **Change** to change the variable name; only an underscore is allowed instead of a space or dash (see Screen 2.6).

A new variable "Income_Replace" was created using *series mean* and ready to be used for further analysis. Go back to the variable view window; the new generated variable is located at the last variable (see Screen 2.7).

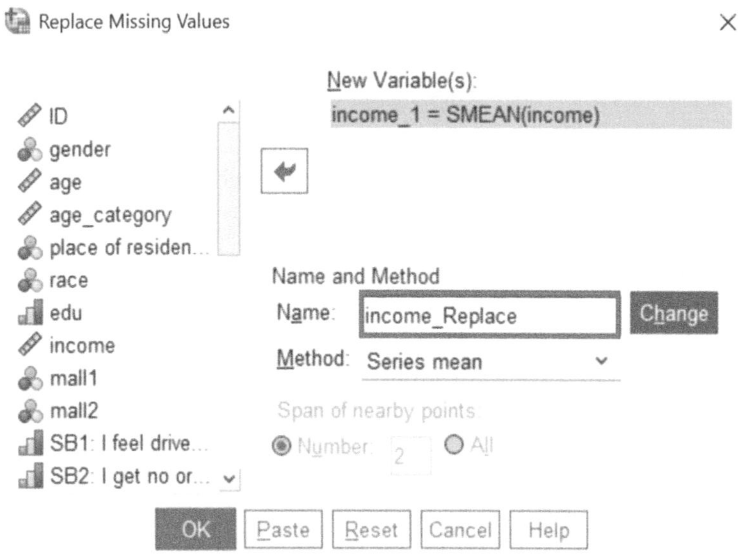

Screen 2.6 Change a new variable name

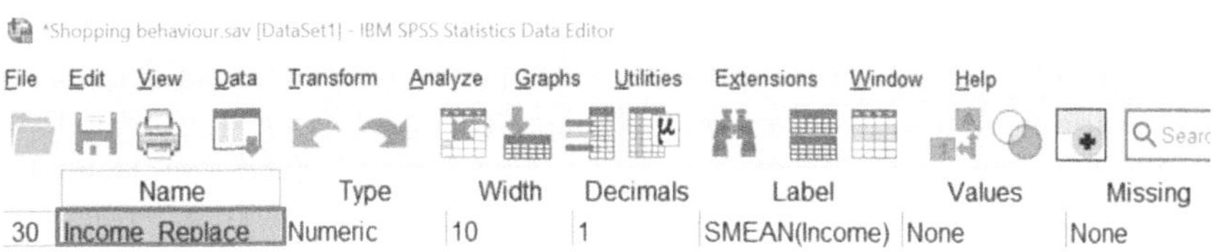

Screen 2.7 Variable view window

There are **five** methods listed in the drop-down list:

1. Series mean is a method that is frequently used for interval or ratio scale data; the missing values will be replaced by the mean score of the particular variable.
2. Mean of nearby points is for replacing missing values by the mean of nearby cases/respondents.
3. Median of nearby points is for replacing missing values by the median of nearby cases/respondents.
4. Linear interpolation is for replacing missing values by two nearby means.
5. Linear trend at the point is for replacing missing values by predicting the variable trend from first case to last case, whether increasing or decreasing.

What if there is a missing value in a categorical data?

Using mode or create a different level for the missing value.

2.2.2 Recode variables

There are two options of recoding a variable: recode into same variables and recode into different variables. The option to recode into same variables means the system will auto-replace the recoded value of the current variable. In other words, the original variable will not be kept and will be replaced by a new value.

Therefore, we frequently choose the option to recode into different variables. By doing so, we can retain the original variable and instruct the system to create a different (new) variable with the recoded values.

We need to recode a variable in the following conditions:

- Existence of a negatively worded statement in the scale,
- Combining different groups into the same group, and
- Simply creating a dummy variable with two groups only or, alternatively, using the Transform-Create dummy variable.

2.2.3 Example of reverse coding a negatively worded statement

It is important to do reverse coding before computing the summation score for a variable that is measured by multiple questions. All questions need to be in the same direction so that the summation score will be correct. Using the shopping behaviour sample data, there are 15 questions with a 5-point rating scale. Generally, the questions are related to positive behaviours towards shopping, except for questions 2 and 3.

Example of a negative statement in question 2: "**Statement b2_sb**: I get no or little pleasure from shopping."

Thus, there is a need to recode the variable so that all the questions are in the same direction. There are two options for recoding an item:

i. Recode into the same variable: The existing variable will be replaced by the recoded results.
ii. Recode into a different variable: A renamed variable will be added as a new variable.

The following is an example of recoding into a different variable so that we can still trace the original data in case of the need for descriptive analysis later on.

1. On the main menu, choose **Transform** and then select **Recode into Different Variables** (see Screen 2.8).

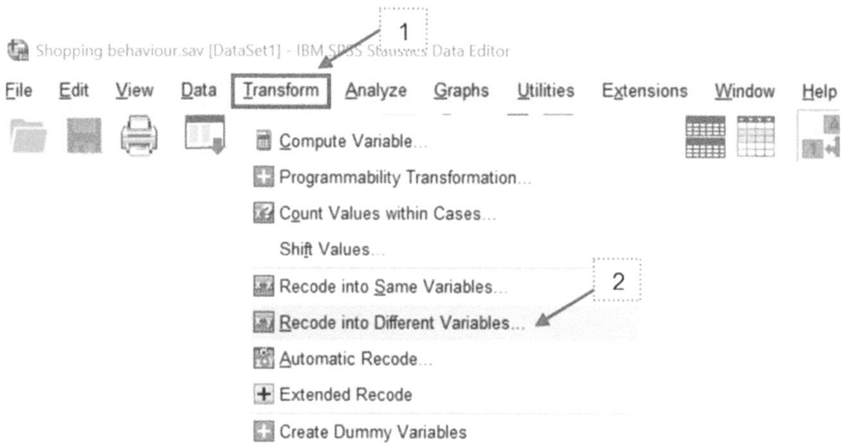

Screen 2.8 Transform and recode

This opens up Screen 2.9, **Recode into Different Variables**.

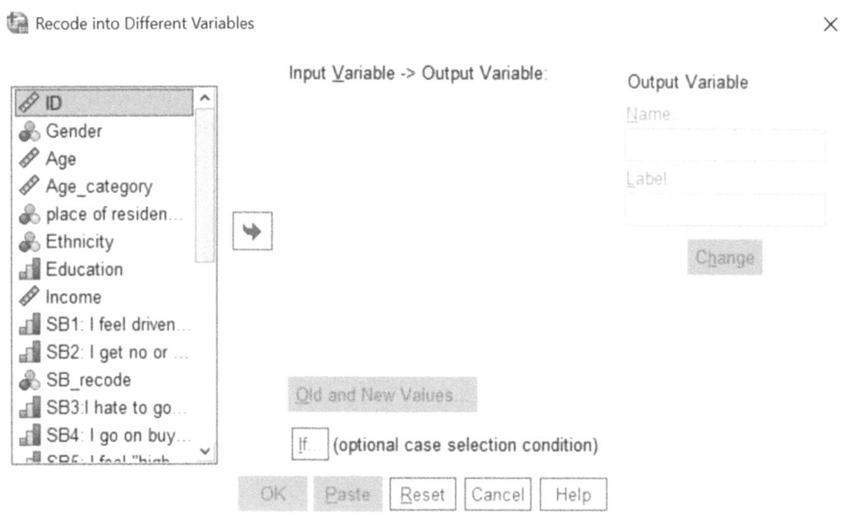

Screen 2.9 Recode into different variables

2. Move the negative statement **b2_sb** into the **Numeric Variable Output Variable** box. Insert name for the **Output Variable**, for example, in Screen 2.10, **SB2_recode** (*no space or dash is allowed for the variable name, but an underscore is fine*).

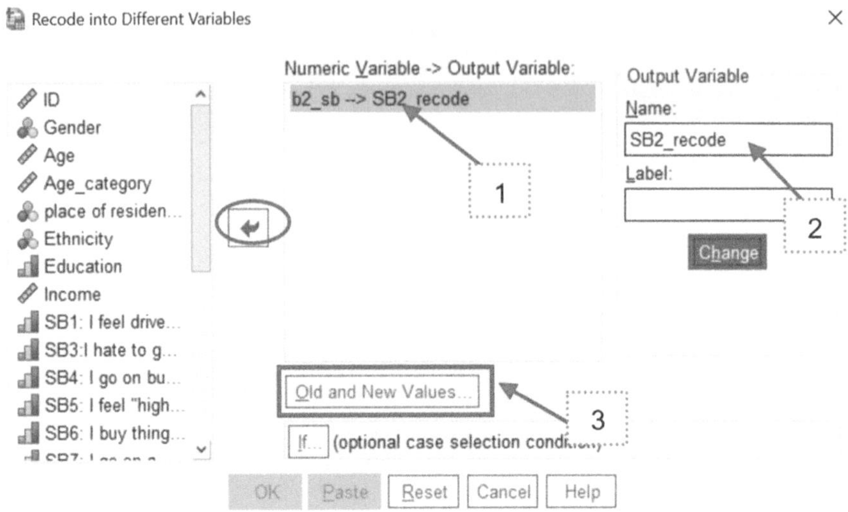

Screen 2.10 Move a variable, rename the output variable, and change old values

Click **Change** to allow the system to create a new variable and rename the new output variable as display in Screen 2.10. Next, click the **Old and New Values** button to enter a new value to replace the old value.

3. In the Recode into Different Variables: Old and New Values, you need to assign a new value when reverse coding, that is, **Old Value 5** to **New Value 1**. Remember to click **Add** to enter the information into the **Old ➔ New** box (Screen 2.11).

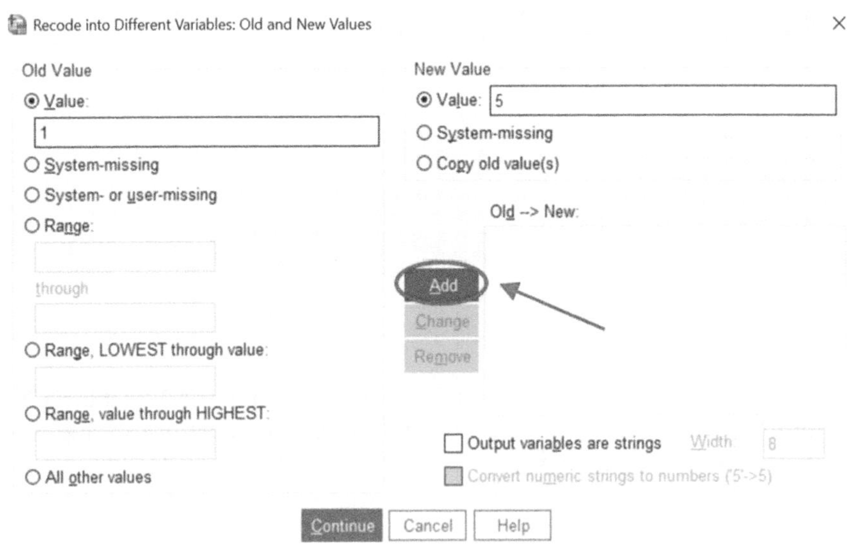

Screen 2.11 Recode into different variables: Old and New Values

4. Repeat the same steps for the rest of the values (4 ➔ 2; 3 ➔ 3; 2 ➔ 4; 1 ➔ 5). Check the working sample in the **Old ➔ New** box in Screen 2.12. After completing the entry and all the values, click **Continue**.

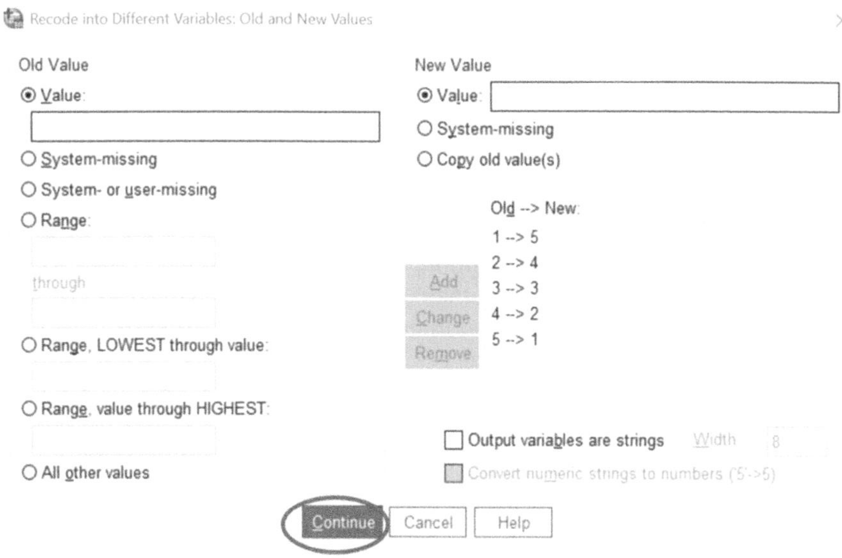

Screen 2.12 Continue to insert old and new values

5. Click **OK** to recode the variable (see Screen 2.13).

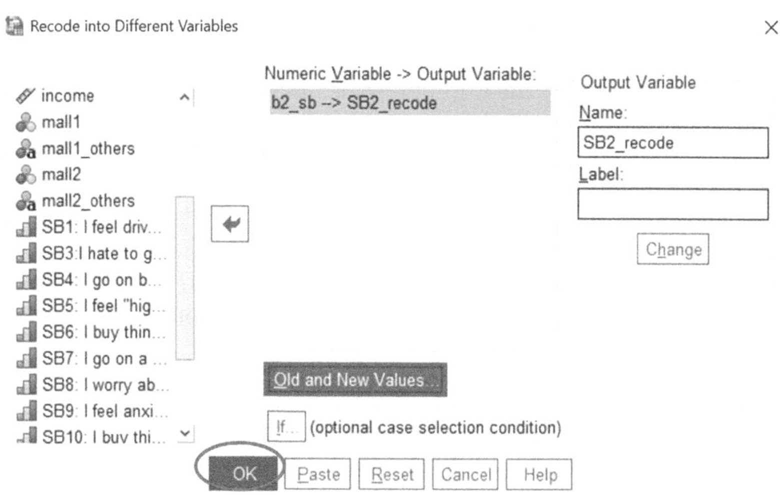

Screen 2.13 Run the test

To check the newly created variable (recoded variable), SB2_recode, go to the last row of the *variable view*. To view the recoded data, go to the *data view*. Always remember to exclude the original negative statement and only include the reverse-coded variable and other positive statements in the reliability test. Similarly, during data computation, for example, to generate the summation score, only include the reverse-coded variable and other positive statements.

To get ready the data before starting to running the suitable data analysis to test our research hypothesis, we need to prepare the data in the required format, which may involve certain mathematical functions (**Transform-Compute**).

2.2.4 Computation

Under the **Transform-Compute option**, a dialogue box similar to the calculator appears. The user can perform the computation by applying the calculator function or mathematical equation in the numeric expression box.

The following section shows the example of how to:

1. Calculate the composite score and
2. Categorize the composite score into two or more categories using a class interval.

Example 1: Calculate the composite score

Working data: Internet banking

Research objective: To identify the relationship between performance expectancy (PE) and behavioural intention (BI) to use internet banking.

To get data ready for correlation analysis, we need the behavioural intention on a continuous scale. In other words, we need to calculate the summation of behavioural intention.

Prior to that, remember to check the **reliability** results to ensure the items are intended to measure the scale. The scale has good internal consistency ($\alpha = 0.92$) and is composed of three items with a 5-point Likert scale. Since there are **no negatively worded** items in this scale, we can proceed to data computation now. If negative statements do exist, recode the items before they can be used to do data computation, such as summation score.

Given that, the behavioural intention was measured by 3 items in the 5-point Likert scale. The possible score ranges from 3 points to 15 points.

The SPSS working steps as follow.

1. Click **Transform** and select **Compute Variable** (see Screen 2.14).

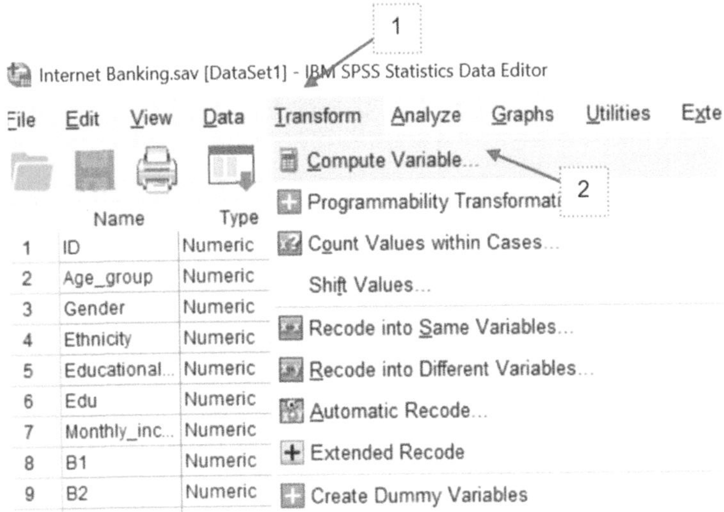

Screen 2.14 Compute Variable under Transform

The dialogue box **Compute Variable** opens (see Screen 2.15). Rename the target variable, and a new variable will be created based on the computation information to be entered. The middle of the screen shows the view of a calculator and mathematical functions on the right.

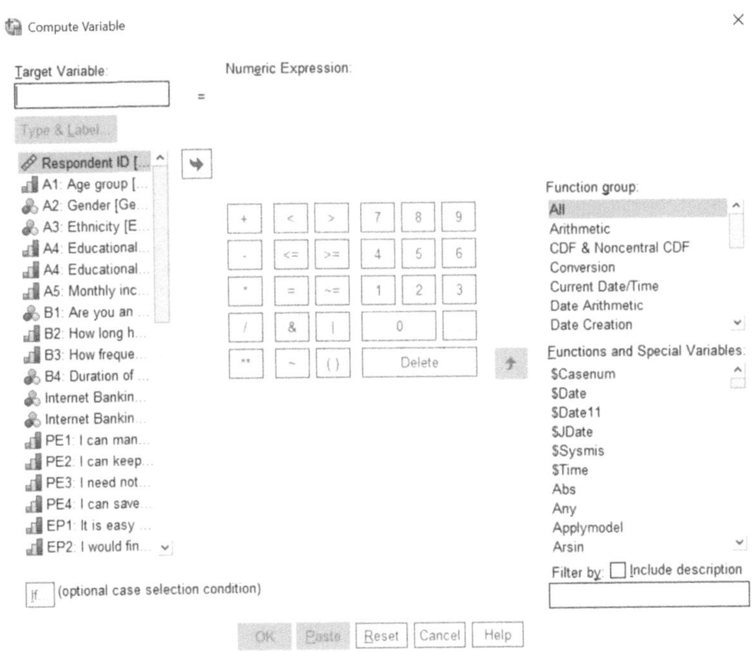

Screen 2.15 Compute Variable window

2. Enter a new variable name **Sum_BT** into the Target Variable box. Scroll down and look for items BT1 to BT3. Move **BT1** into the **Numeric Expression** box, and click the addition function **(+)**. Then, move **BT2**, and click **(+)**, and lastly move the **BT3** (see Screens 2.16–2.17).

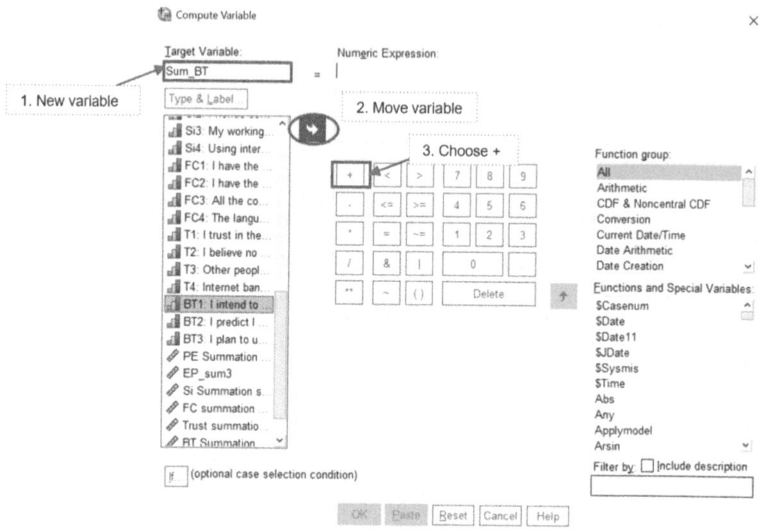

Screen 2.16 Enter name for the targeted variable; insert variable and functions

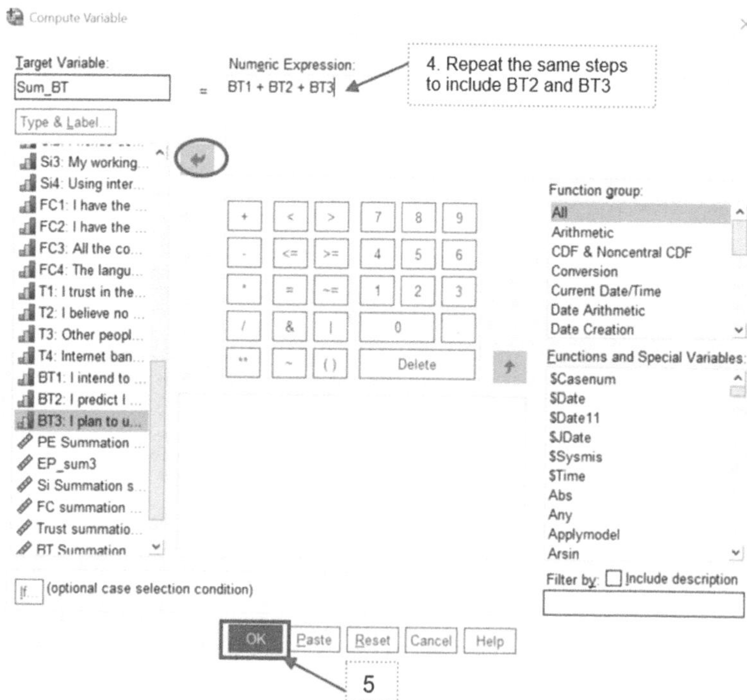

Screen 2.17 Complete the variable selection

> **Tips to success**
>
> If there is a negative statement, insert the **recoded variable** (*instead of the original variable*) into the Numeric Expression.

3. Check all of the Numeric Expression items and click **OK** to run the procedure.

Check at variable view

The additional one variable, **Sum_BT**, is created in the last row. There will be no output produced (see Screen 2.18).

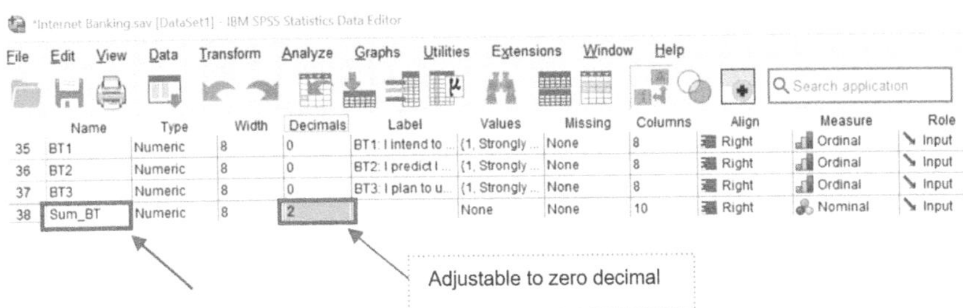

Screen 2.18 Check the new variable at variable view

Check at data view

The additional column, **Sum_BT**, is created to record the generated summation values (see Screen 2.19).

THE ALTERNATIVE METHOD

Calculate the composite score using the methods of the mathematical equation.

After reading the following table, you will know what should be entered in the numeric expression box using different calculation methods. The mathematical equation is the same as the function we use in Excel.

Table 2.1 Sample of methods to generate a composite

	Calculator function	*Mathematical equation*
Composite score	BT1+BT2+BT3	SUM (BT1, BT2, BT3)

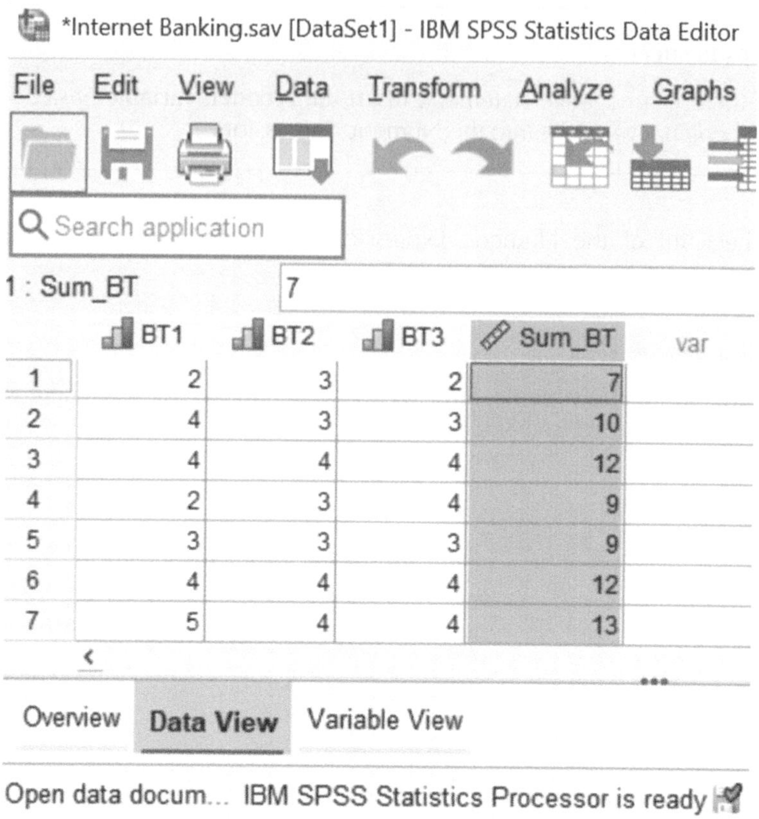

Screen 2.19 View summation values at data view

Continue from Screen 2.15 to generate the summation score. Go to **Function group**, select the option **All**, and move to next the box, **Functions and Special Variables**. Choose **Sum**, move the variables one by one to replace the question mark, and separate the variables with commas (see Screens 2.20–2.21).

Example 2: Categorize the composite score into two or more categories using a class interval

Generally, we can categorize composite scores or summation scores as a different variable based on a theory or equal groups. For example, to find out the trend of respondents' body mass index (BMI) ranges, at first, we need to collect the respondent's height and weight. After that, tabulate the BMI index using the formula, BMI = weight (kg)/height (m)^2 from

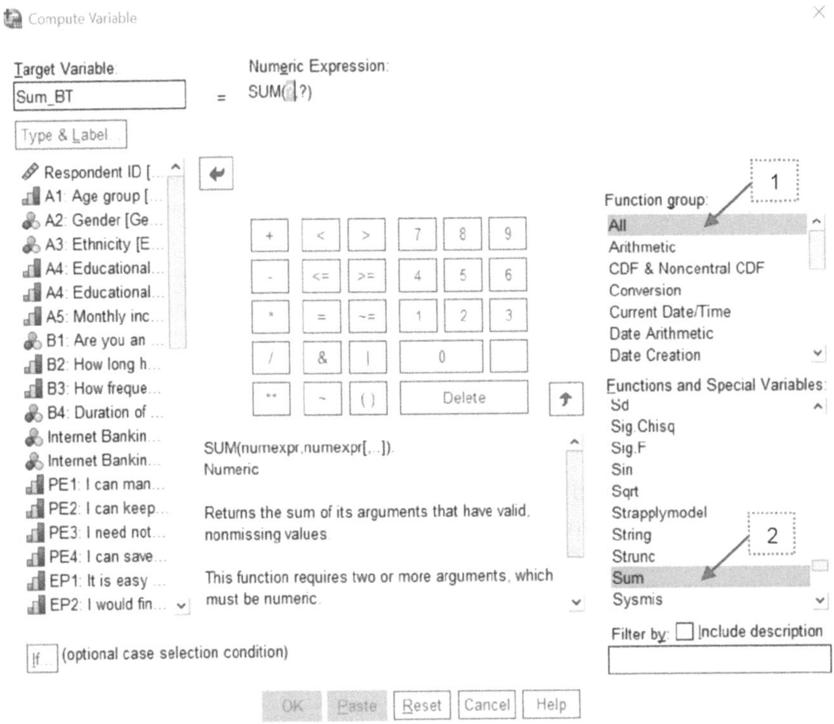

Screen 2.20 Compute a variable using function group

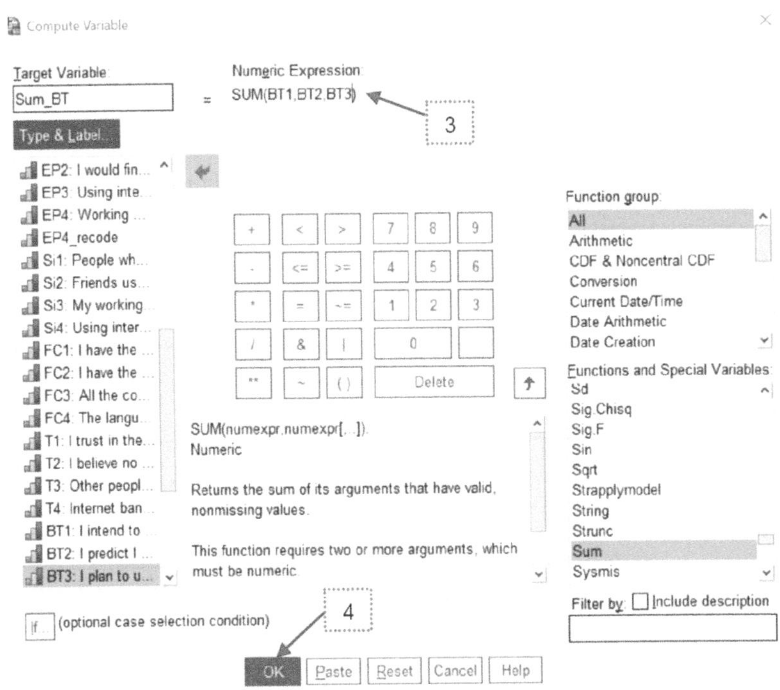

Screen 2.21 Insert variables using function group

WHO. Remember, always to use the function of Transform – Recode into a different variable so that SPSS can still keep the original variable. From these BMI indices, we can now further categorize the data based on a different range of BMI. This is how we do categorization based on a theoretical basis.

However, in certain cases, when we do not have a theoretical background or an index to refer, the alternative way is to categorize data using the equal range method/class interval.

Working sample using Internet banking data

Objectives: Categorized behavioural intention into three (3) groups (low, moderate, and high) using equal range method/class interval.

Firstly, run a frequency test to know the lowest scale value, highest scale value, and the range.

Next, calculate class width and create three (3) groups of class intervals in a table.

1. Determine the class width of the class interval based on the scale (behavioural intention).
   Class width (CI) = (HSV – LSV)/K
   HSV = highest scale value
   LSV = lowest scale value
   K = number of categories
   Class width = (15–3)/3
   = (12)/3
   = 4
2. Create the class interval (CI) using class width and refer to the formula below.

Class interval (CI) = Upper class boundary (UCB) – Lower class boundary (LCB)

Table 2.2 Determine class interval

Class no.	*Class width*	*Class*	*Class boundary*		*Class interval (exact limit)*
			Lower class boundary (LCB)	*Upper class boundary (UCB)*	
1	4	3.00–7.00	3.00	7.00	3.00–7.00
2	4	7.00–11.00	7.01	10.99	7.01–10.99
3	4	11.00–15.00	11.00	15.00	11.00–15.00

SPSS WORKING STEPS

1. Go to the main menu, then choose **Transform – Recode into different variable**. Move the **BT_Summation** into **Recode into different variable** (See Screen 2.22).

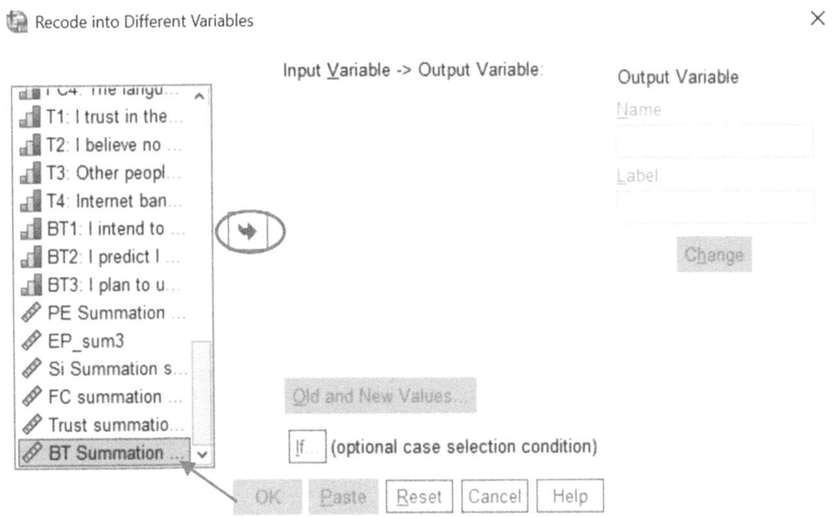

Screen 2.22 Move the input variable

2. Under the **Output Variable**, enter **BT_CAT** in the **Name** box and **BT_categories** in the **Label** box.

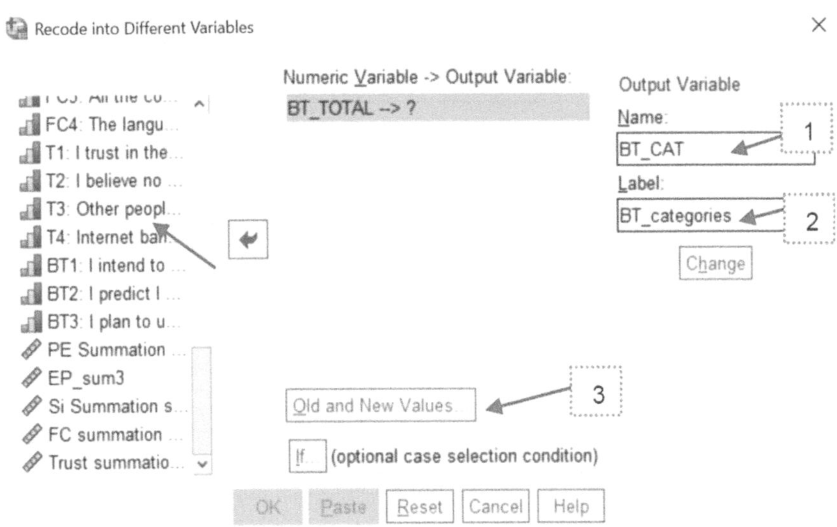

Screen 2.23 Rename the output variable and label

3. Click **Old and New Values**, choose **Range**, then start to enter the class interval value and a **New Value**.

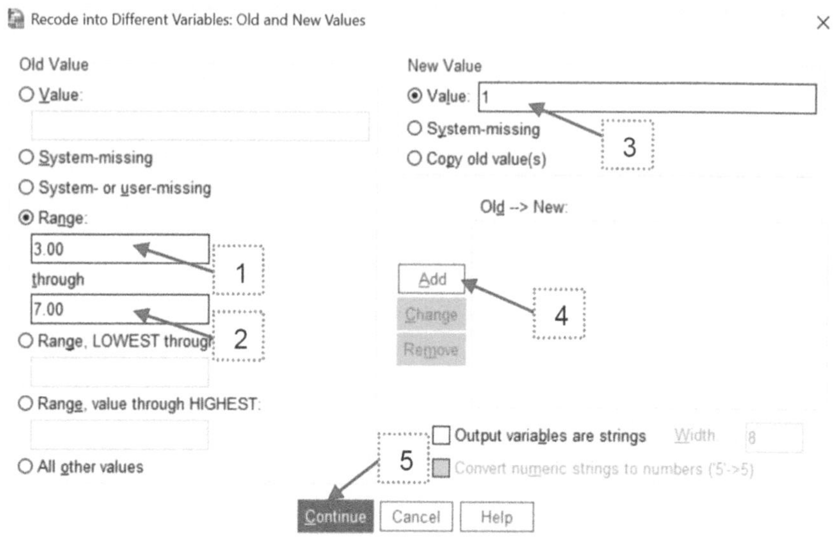

Screen 2.24 Recode variable from old to new values

Click **Add** to add, and repeat for each **Range**, and then repeat the step to add another range.

Range (Class interval)	*Value (New value)*
3.00–7.00	1
7.01–10.99	2
11.00–15.00	3

Click **Continue** to go back to the main dialogue of **Recode into different variable**.

4. Click **Change** and **OK** to run the procedure.

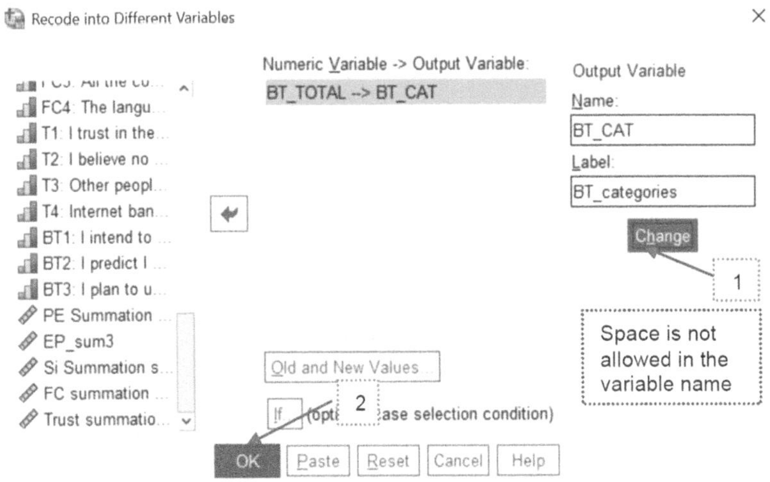

Screen 2.25 Recode into different variable – change

SPSS OUTPUTS

Check at variable view The additional one variable **BT_categories** is created at the last row.

Remember to change the **Value Label** at variable view so that you may know which category that the respondents belong to.

Insert Value: 1, Label: Low; click Add to insert the information. Then repeat the next values: Value: 2, Label: Moderate; Value: 3, Label: High.

After entering all the information, click **OK** to change the **Value Label**. Furthermore, change the **Decimals** from two to zero or adjust accordingly based on the numeric nature of the variable.

Check at value view

The additional column **BT_categories** is created to show the values of each category.

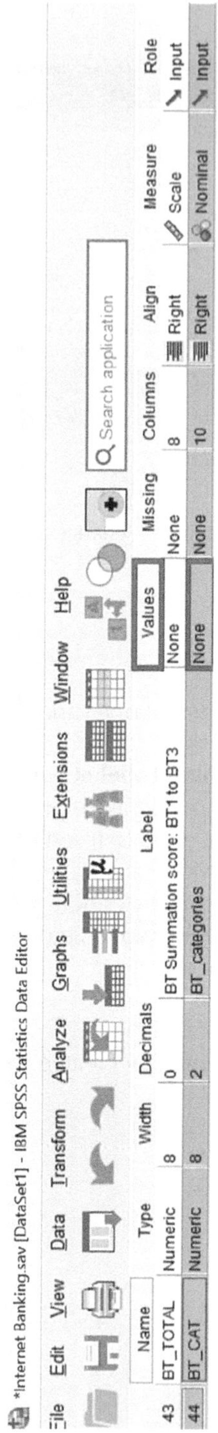

Screen 2.26 Variable view – last variable

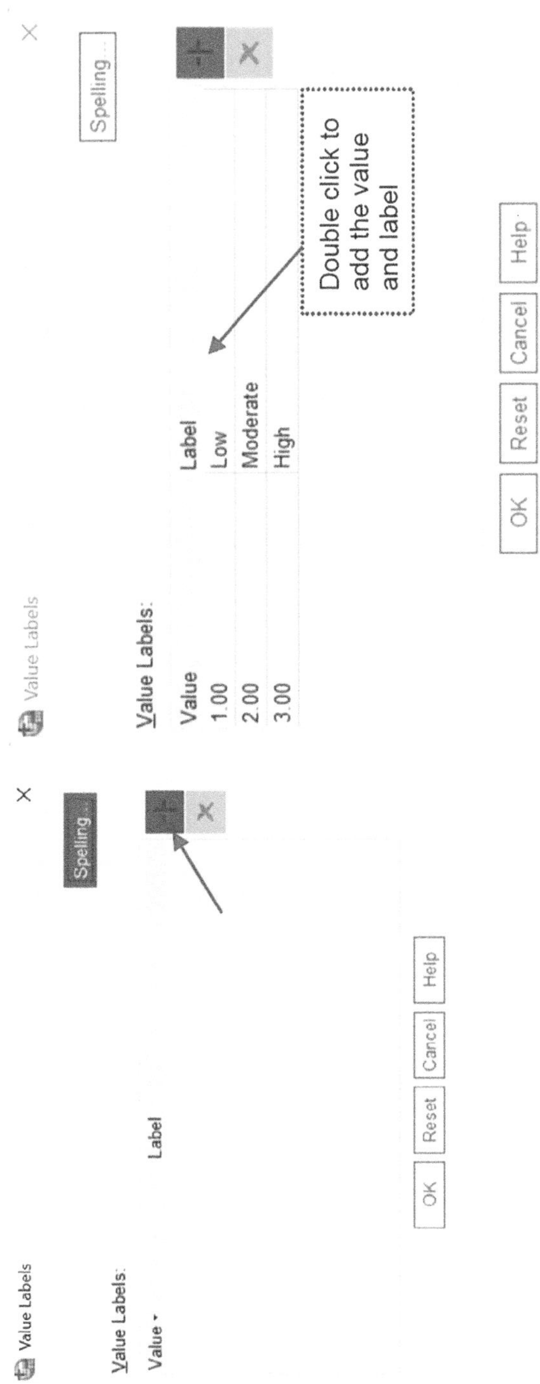

Screen 2.27 Variable view – change value label and decimals

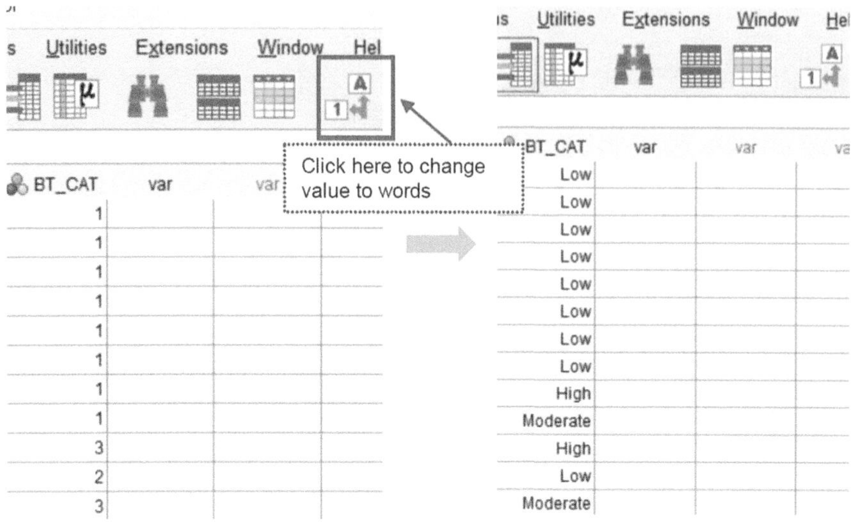

Screen 2.28 Data view

3 Reliability test

Learning outcomes

At the end of this section, participants will be able to:

- Recognise the purposes and requirements of a reliability test,
- Run a reliability test step by step using SPSS,
- Interpret the reliability test outputs, and
- Report a reliability test in APA format.

3.1 Definition of reliability test

Reliability indicates how well the scale produces consistent results without bias; it confirms the scale is a consistent measurement across time (stability), including all items in the specific scale for checking if they are measuring the same construct (internal consistency).

Types of reliability test

1. ***Test-retest***
   The repeatability of test scores measured at a different time. It can check the stability of the characteristic or construct of the scale. Hence, a researcher sends the same questionnaire to the same person twice.
2. ***Alternate or parallel form***
   The test estimates how well the items were measured with the same information. A researcher intends to use more than one form to measure the same information alternately.
3. ***Inter-rater***
   The test can check whether the test scores are consistent, although scored by two or more raters.
4. ***Internal consistency***
   The test is measured by Cronbach's alpha (α), which indicates if the items are measuring the same thing. It is calculated using the correlation of all items in the same test.

DOI: 10.4324/9781003673903-3

When do we need reliability?

☒ The questions in the demographic background
☒ Open-ended questions (check for validity instead of reliability)
☑ After a pilot test of a quantitative study
☑ Adopted/adapted scale
☑ Scale development

Things to know before running a reliability test

1. Remember to recode the negative statement. Exclude the negative statement and include the recoded statements.
2. Include all the questions or statements that measure the selected variable.
3. Run another reliability test for a different variable.

The Cronbach's alpha value ranges from 0 to 1; any value closer to 1 is better. The rule of thumb to interpret the level of acceptance of alpha value (***α***) by George and Mallery (2024) is as follows:

Cronbach's alpha (α)	*Internal consistency*
< 0.5	Unacceptable
$0.51 \leq \alpha < 0.60$	Poor
$0.61 \leq \alpha < 0.70$	Questionable
$0.71 \leq \alpha < 0.80$	Acceptable
$0.81 \leq \alpha < 0.90$	Good
> 0.9	Excellent

Working example

The working data is from Yeoh and Chan (2011) for the published article titled "Internet banking adoption in Kuala Lumpur: an application of the UTAUT model."

For example, effort expectancy was adapted from a previous study (Venkatesh et al., 2003). Run a reliability test for this scale variable, and remember, do not combine with other selected scale variables in the study. Each scale variable needs a separate reliability test.

Let's check the following session for SPSS steps on how to run the reliability test.

3.2 Steps to run reliability test

1. On the menu bar, select **File**, **Open**, and **Data** (see Screen 3.1). The Open dialogue box appears.

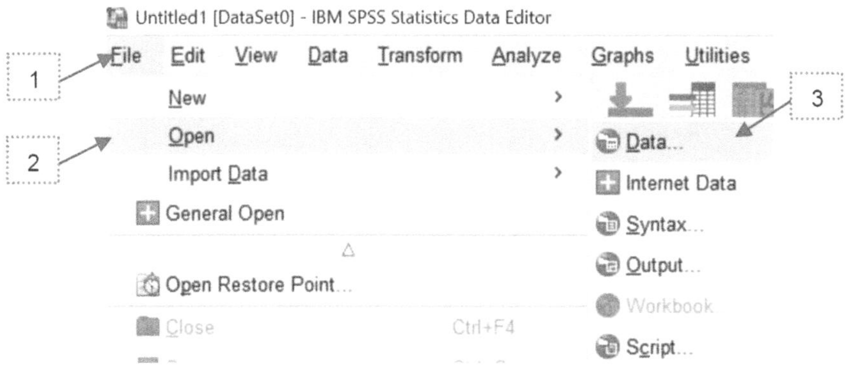

Screen 3.1 Data editor

2. In the sample data folder, select the **Internet Banking** data file and click **Open** to open the data file (see Screen 3.2).

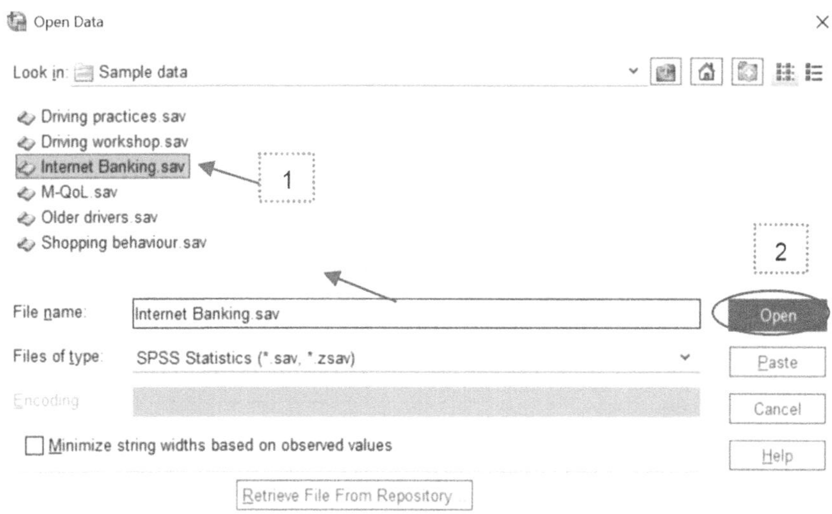

Screen 3.2 Open data

3. On the main menu, click **Analyze**, go to **Scale,** and click **Reliability Analysis** (see Screen 3.3).

Screen 3.3 Main menu: Choose reliability test

The **Reliability Analysis** dialogue box appears (see Screen 3.4).

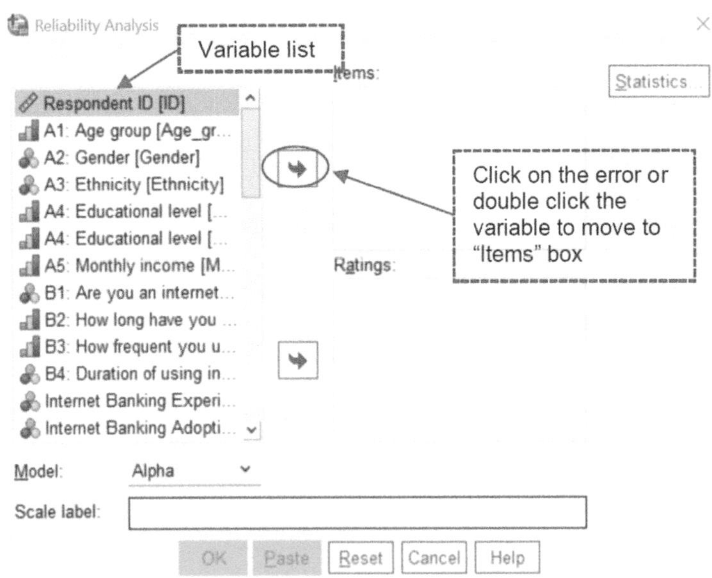

Screen 3.4 Reliability analysis

4. In the variable list, scroll down and search for **EP1:EP3** and move them into the **Items** box (see Screen 3.5).

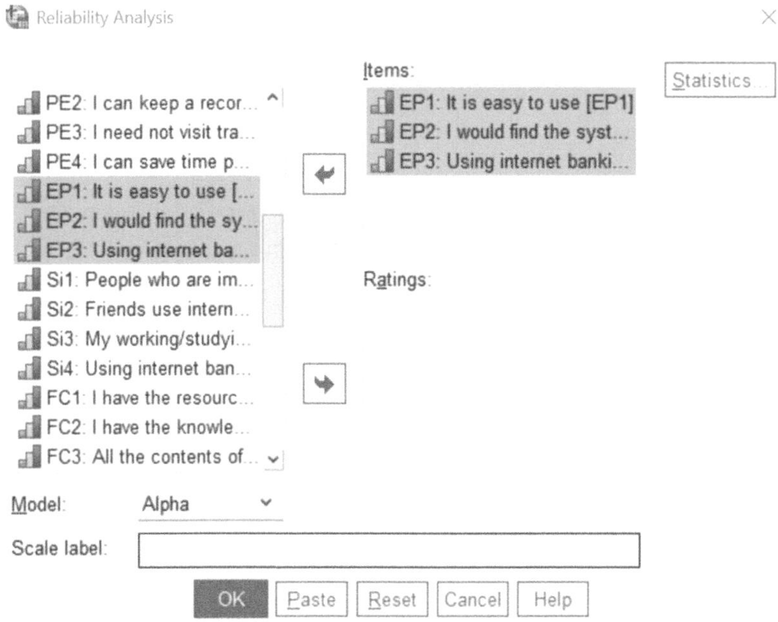

Screen 3.5 Reliability test: move variables to the items box

5. Click **Statistics,** and this opens up the **Reliability Analysis: Statistics** dialogue box (see Screens 3.6–3.7).

   Under **Descriptives for**, select Item, Scale, and Scale if the item is deleted. Next, click **Continue** to go back to the **Reliability Analysis** dialogue box.

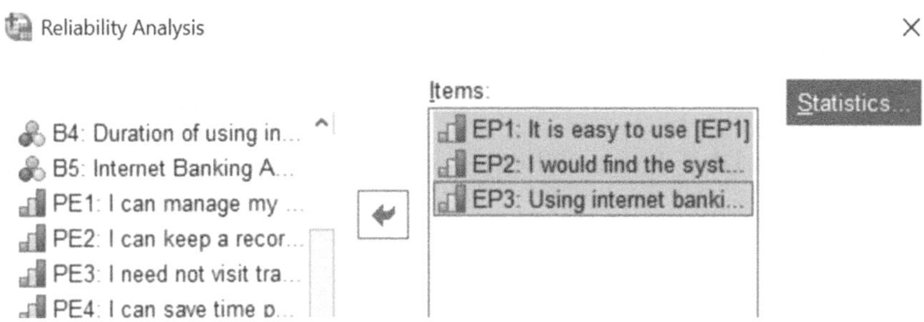

Screen 3.6 Reliability test: Statistics

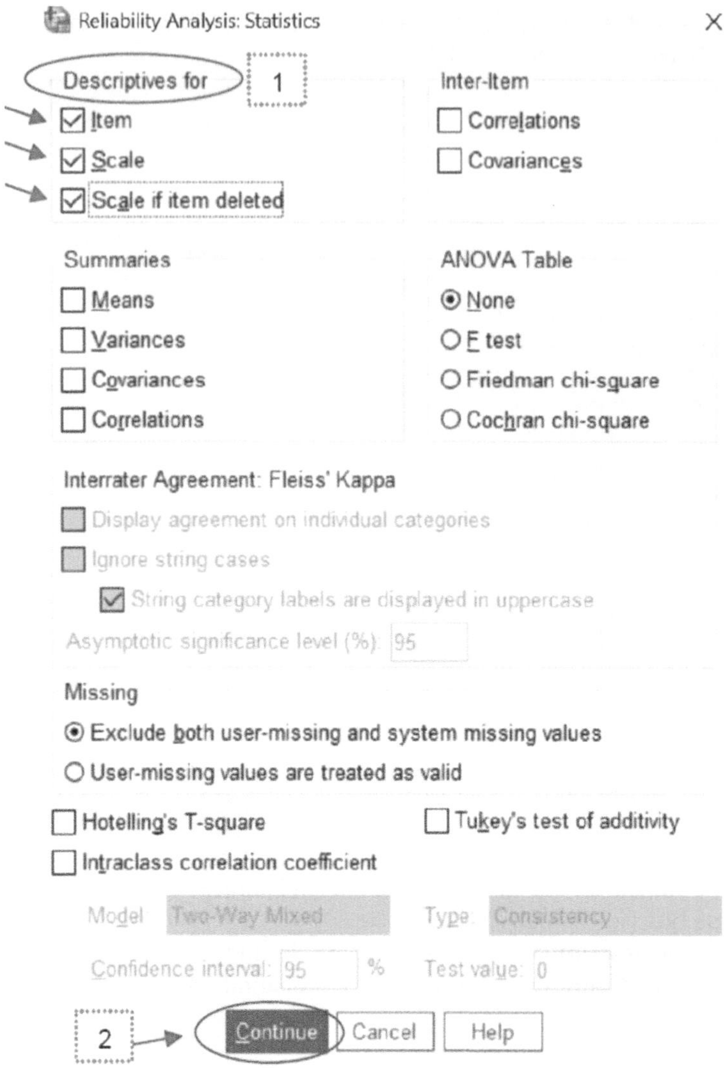

Screen 3.7 Reliability test: Statistics – Descriptives for

6. In the **Scale label**, key in "Effort Expectancy" and click **OK** to run the test (see Screen 3.8).

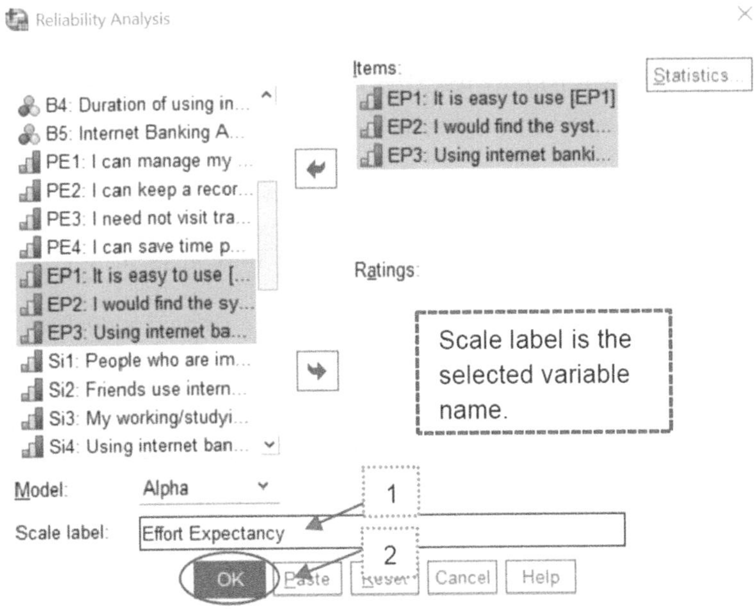

Screen 3.8 Reliability test: Statistics – Effort Expectancy

SPSS outputs

Reliability Statistics

Cronbach's Alpha	N of Items
.800	3

Item Statistics

	Mean	Std. Deviation	N
EP1: It is easy to use	3.46	1.036	200
EP2: I would find the system to be flexible to interact with	3.34	.920	200
EP3: Using internet banking save me a lot of time	3.57	1.035	200

Item-Total Statistics

	Scale Mean if Item Deleted	Scale Variance if Item Deleted	Corrected Item-Total Correlation	Cronbach's Alpha if Item Deleted
EP1: It is easy to use	6.90	2.894	.693	.675
EP2: I would find the system to be flexible to interact with	7.02	3.447	.620	.756
EP3: Using internet banking save me a lot of time	6.79	3.061	.629	.745

Scale Statistics

Mean	Variance	Std. Deviation	N of Items
10.36	6.411	2.532	3

Screen 3.9 Outputs of the reliability test

3.3 Interpret the reliability test SPSS output

1. Identify Cronbach's alpha value from the Reliability Statistics (see Screen 3.9).

   The possible score of the alpha value is between 0 and 1. The corresponding Cronbach's alpha is **.80**. It is a relatively high alpha, and the scale is considered to have acceptable consistency. Thus, this scale is considered reliable.
2. Check the values of Cronbach's alpha if the item is deleted under Item – Total Statistics.

The range of alpha value under Cronbach's alpha if an item is deleted ranges from .675 to .756.

If the **EP1** was deleted and only the EP2 and EP3 are present, the alpha value will change to **.675.**

Similarly, if EP2 is deleted and the other items remain, the alpha value will become **.756**.

If we compared the obtained Cronbach's alpha, it is **.80**; any deletion of the item will lower the overall Cronbach's alpha value. Thus, we suggest retaining all three items in effort expectancy.

The effort expectancy has an acceptable internal consistency of the overall scale scores (α = .80).

- Let's say, a deletion of the item could increase Cronbach's alpha value. The student needs to evaluate and consider removing the item. Hence, another reliability test needs to be performed if there is item deletion. All of the details and results need to be recorded, justified, and reported in the report or thesis.

Tips to succeed in a reliability test

If the scale has any **negative statements**, remember to **recode** it. This groundwork should be done after you check the missing data.

Make sure you enter the recoded negative statement during the reliability test. For instance, EP3 is a negative statement. Recode into a different variable and then enter EP1, EP2, and the **recoded EP3** into the reliability test. Do not include the original data of EP3.

Where should I include the reliability test result in my thesis?

1. Present pilot test results in Chapter 3: Research Methodology.
2. Present the reliability test after the final data collection in the chapter of results and discussion, usually in Chapter 4 of the thesis.

3.4 Writing the results in APA format

A reliability test was conducted to check the internal consistency of effort expectancy using Cronbach's alpha. The initial reliability of the scale was found to be (α = .80). Cronbach's alpha is considered acceptable based on the rule of thumb by George and Mallery (2024). A closer inspection of the itemized statistics shows that there is no additional increase in Cronbach's alpha if the item is deleted (see Table 3.1). Hence, all three items were retained with acceptable internal consistency.

Table 3.1 Reliability test of effort expectancy (EP)

	Scale mean if item deleted	*Scale variance if item deleted*	*Corrected item – total correlation*	*Cronbach's alpha if item deleted*
EP1: It is easy to use.	6.90	2.894	.693	.675
EP2: I would find the system to be flexible to interact with.	7.02	3.447	.620	.756
EP3: Using internet banking saves me a lot of time.	6.79	3.061	.629	.745

Notes: Cronbach's alpha (α) = 0.8; *Var* = 6.41; *SD* = 2.53; number of items = 3

The summary table of all the selected variables can be presented in a table (see Table 3.2).

Table 3.2 Summary table of the reliability test for the selected variables

Variables	*Cronbach's alpha* (α)	*No. of items*	*Variance*	*Standard deviation*
IV1	.80	3	6.41	2.53
IV2				
IV3				
IV4				
DV				

Note: Remember to include all items of each variable (except antecedent variables). Report the internal consistency for dichotomous scale items using the Kuder–Richardson Formula 20 (KR-20). The KR-20 value will be generated by SPSS using the same reliability test procedure.

4 Descriptive analysis

Learning outcomes

At the end of this section, participants will be able to:

- Recognise the various types of descriptive statistics,
- Perform descriptive analysis using SPSS,
- Know which statistic to report based on different scales of measurement,
- Generate and display data in a graph, and
- Present descriptive statistics table and results in APA reporting style.

4.1 Introduction of descriptive analysis

When do you need a descriptive analysis? If a researcher is conducting a descriptive study and/or wants to explore the background of respondents, to report item analysis for each question (frequency distribution), or to display collected data in a graph.

In terms of graph displays, data in both interval and ratio can be displayed in a histogram or a frequency polygon.

Nominal and ordinal scale data are appropriate to display by pie charts and bar charts, respectively.

Guideline to report the different types of statistics under the measure of central tendency and dispersion for different scales was summarised in Table 4.1.

4.2 To generate output for categorical data

If the aim is to present the profile or demographic information of the participants/respondents. Take this example, the variables of gender and age groups were selected to generate the descriptive information.

DOI: 10.4324/9781003673903-4

Table 4.1 Types of statistics to be reported for different scales

	The measure of central tendency			*Measure of dispersion*	
Scales	*Mode*	*Median*	*Mean*	*Variance*	*Standard deviation*
Nominal	✓				
Ordinal	✓	✓			
Interval	✓	✓	✓	✓	✓
Ratio	✓	✓	✓	✓	✓

1. Go to the menu bar, select **File**, **Open**, and **Data** (see Screen 4.1).

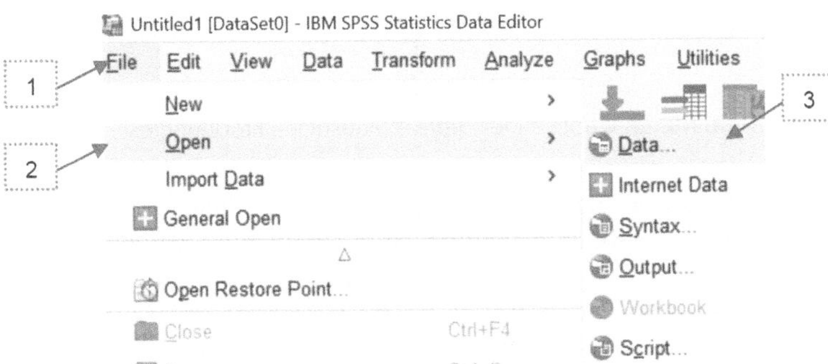

Screen 4.1 Data editor

2. In the sample data SPSS folder, select the **Internet Banking.sav** and click **Open** to open the data file (see Screen 4.2).

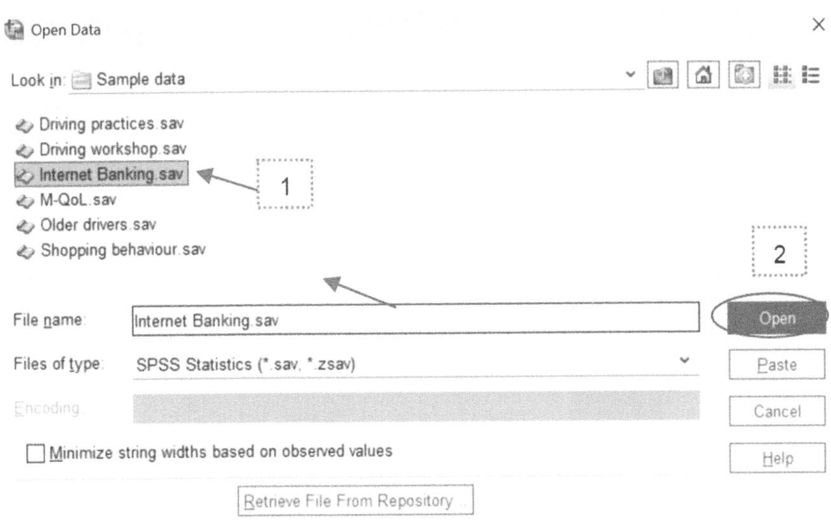

Screen 4.2 Open data

3. On the main menu, click **Analyze,** then go to **Descriptive Statistics**, and select **Frequencies** (see Screen 4.3).

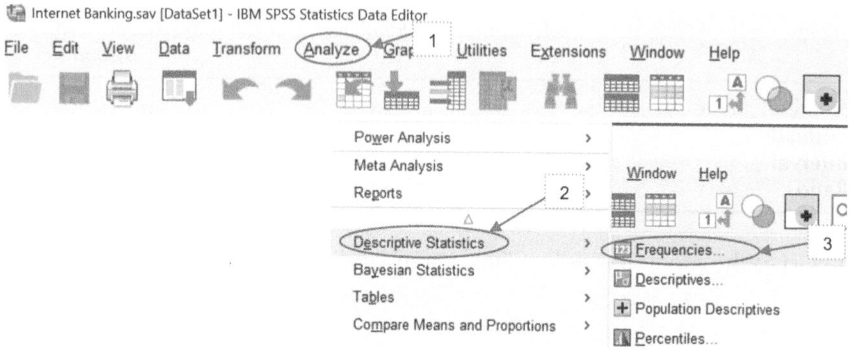

Screen 4.3 Main Menu: Choose Descriptive Statistics – Frequencies

Screen 4.4 is the dialogue box of Frequencies. Tick the option to create APA-style tables, which is optional. It is useful to view what information needs to be presented in the report. Usually, the user can create their own table and move the needed data because the font and colour formatting from the output may not fit the preferences of everyone.

From the variable list box, select and move the variable **Age group, Gender** into the **Variable(s).**

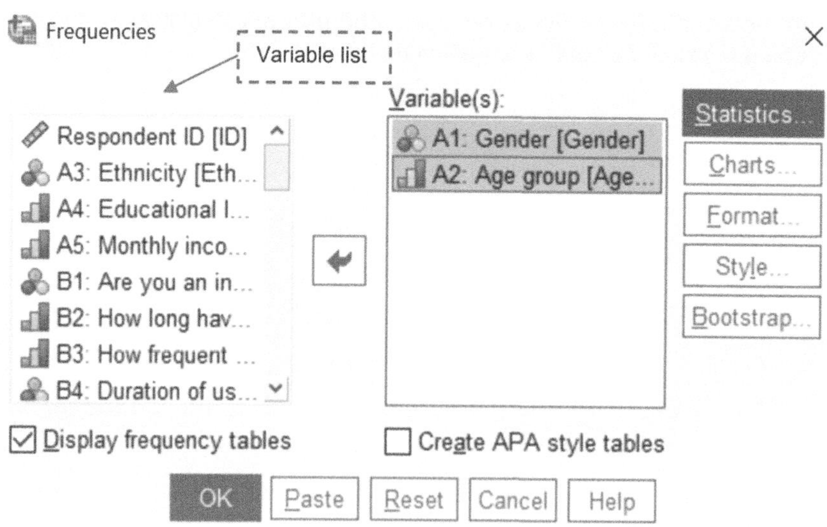

Screen 4.4 Frequencies

4. In Screen 4.4, choose the function of **Statistics,** and Screen 4.5 appears. Under the **Central Tendency**, tick the **Mode** box and click **Continue**.

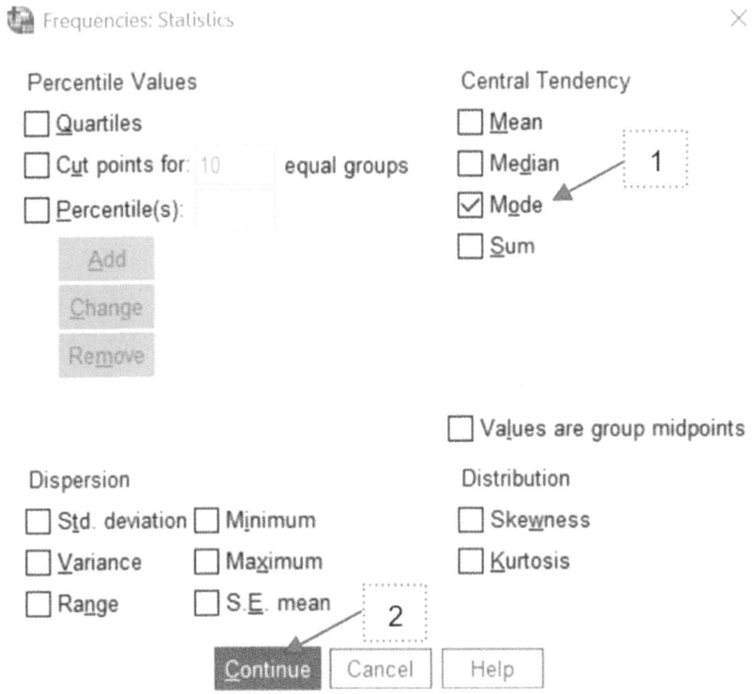

Screen 4.5 Statistics: Frequencies

5. This is the dialogue box of main **Frequencies**. Click **Charts** to select the type of chart to be generated (see Screen 4.6).

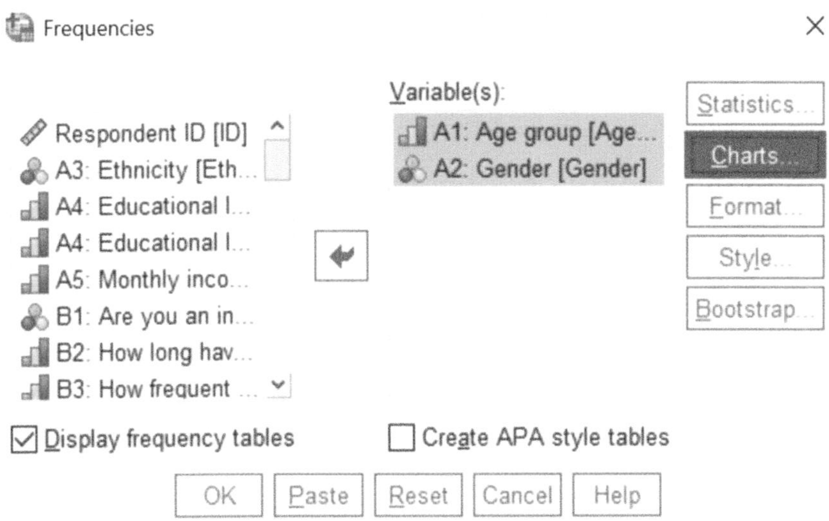

Screen 4.6 Statistics: Charts

6. Under **Chart Type**, select **Bar charts** (alternatively, pie chart), and then click **Percentages** under **Chart Values** (see Screen 4.7). Thus, the system will generate bar charts to show percentages only. Click **Continue** to go back to the main **Descriptive: Frequencies**.

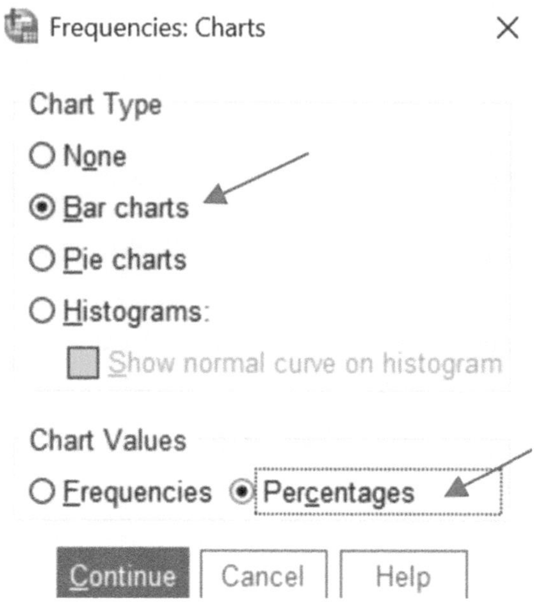

Screen 4.7 Chart type and values

7. Click **OK** to perform the test (see Screen 4.8).

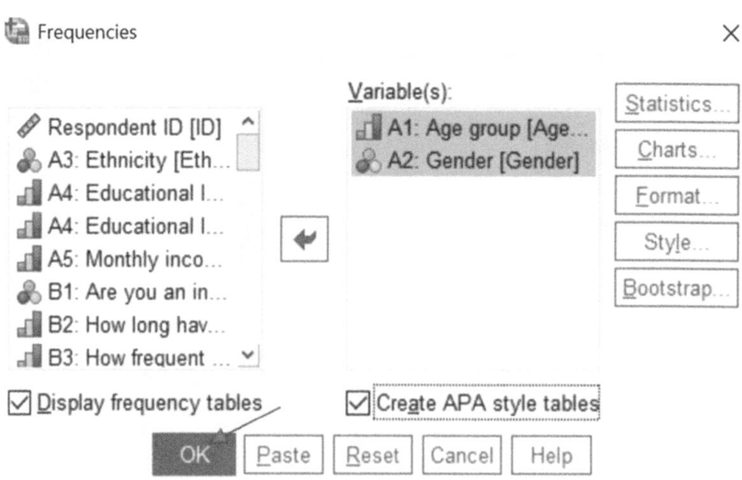

Screen 4.8 Chart type and values

4.2.1 SPSS outputs and interpretation

Screen 4.9 is the frequency outputs created in APA style by SPSS.

Frequencies

Statistics

		A1: Gender	A2: Age group
N	Valid	200	200
	Missing	0	0
Mode		2	1

Frequency Table

A1: Gender

	N	%
Male	93	46.5%
Female	107	53.5%

A2: Age group

	N	%
21-30	89	44.5%
31-40	62	31.0%
41-50	49	24.5%

Screen 4.9 SPSS outputs

Check the value label at variable view. The coding of gender for female is 2, and the age group of 21 years to 30 years is 1.

Bar charts of gender and age groups are in Screen 4.10. The chart display can be edited by double-clicking on the bar chart in the SPSS output. For example, you can change the chart colour, adjust the font size, insert data labels and percentages, etc.

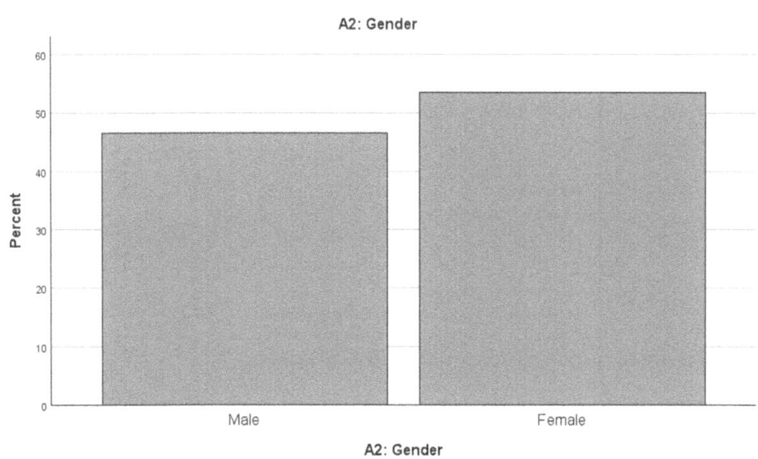

Screen 4.10 Gender by percentages

It is advisable to edit it for a better visual display. As you can see the difference of the visual display in Screen 4.10 and Screen 4.13. To add data labels in the chart, double-click the chart and select **Show Data Labels** under **Elements** to add the percentage values (see Screens 4.11–4.12).

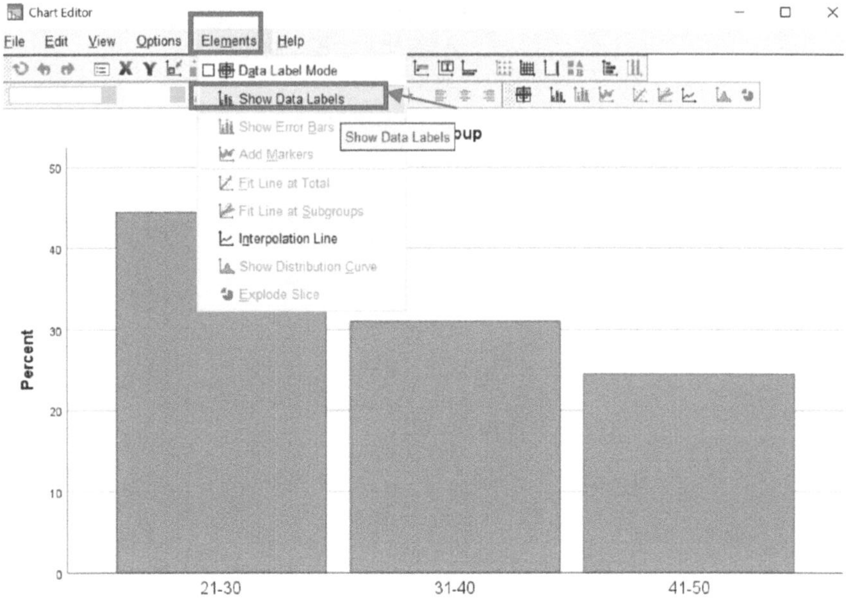

Screen 4.11 Chart editor

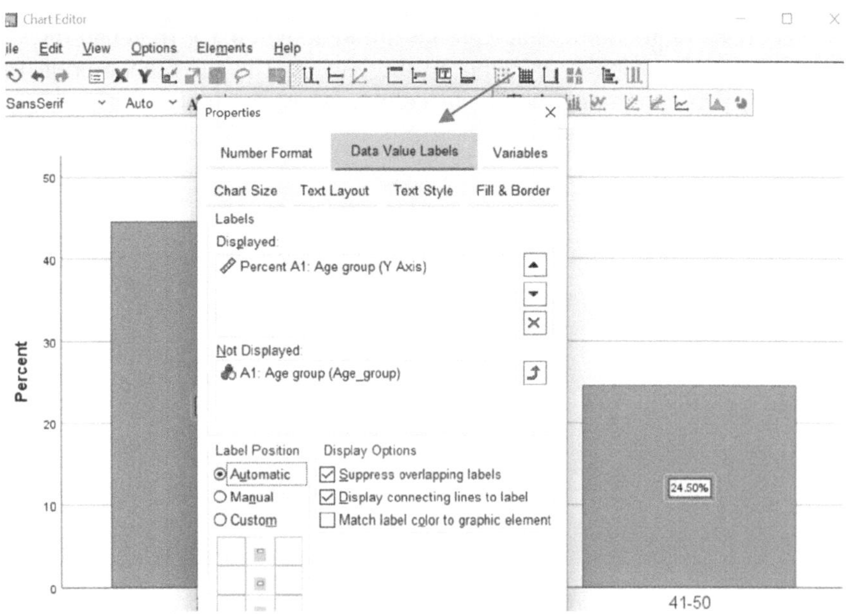

Screen 4.12 Chart editor – Data value labels

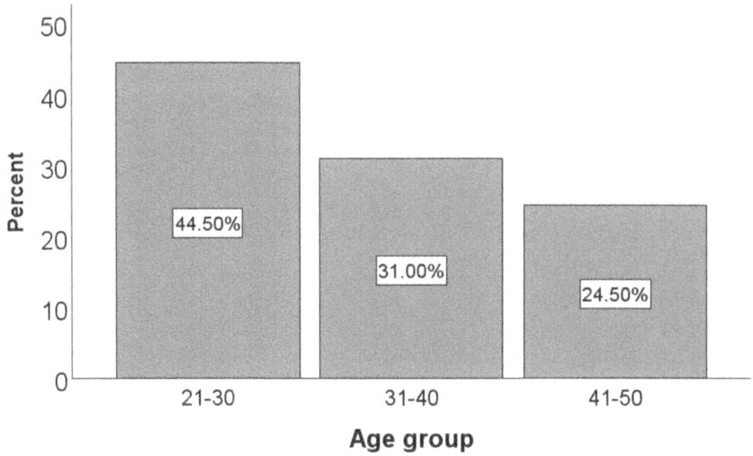

Screen 4.13 Age groups by percentages

4.2.2 Sample of write-up in APA format

A total of 200 complete questionnaires were analysed in this study. More than half of the respondents are female (53.5%), and the remaining are male respondents (46.5%) (see Table 4.2). Generally, the respondents are young adults and stand a larger percentage between 21 and 30 years old (44.5%). This is followed by 31.0% of the respondents in the age category of 31 to 40 years old, and 41 to 50 years old are a minority group (24.5%) in this study.

Table 4.2 Profile of respondents

	Frequency (n)	*Percent (%)*
Gender		
Male	93	46.5
Female	107	53.5
Age group		
21–30	89	44.5
31–40	62	31.0
41–50	49	24.5

4.3 To generate output for continuous data

This section will use the same data file (Internet Banking) to generate the descriptive statistics of the behavioural intention summation score (interval scale).

1. Go to the main menu, click **Analyze,** then go to **Descriptive Statistics**, and select **Frequencies**. Click **Reset** to clear Screen 4.14 (if you had just completed a descriptive analysis just now; otherwise, skip the reset).

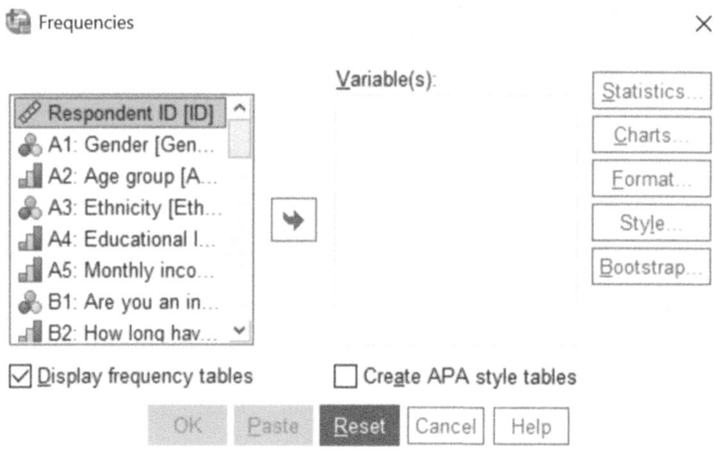

Screen 4.14 Descriptive – Frequencies

2. From the variable list box, select and move the variable **BT Summation** into the **Variable(s)** (see Screen 4.15).

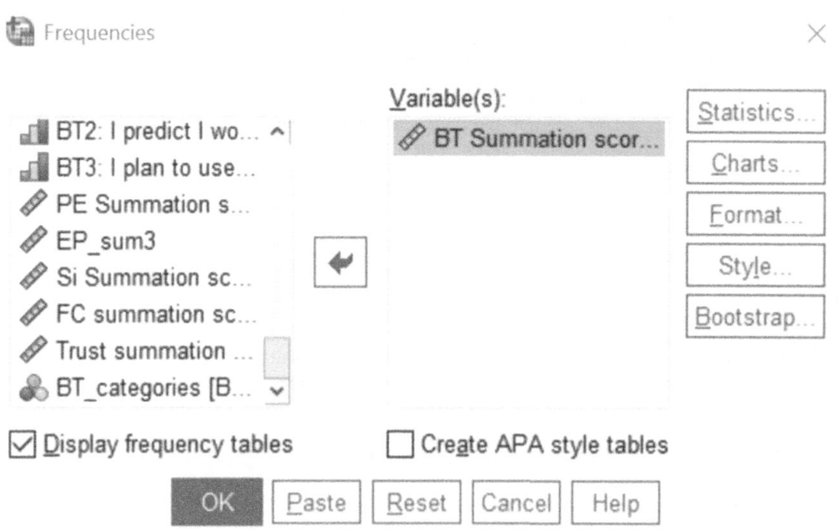

Screen 4.15 Frequencies – Insert variable

3. In Screen 4.15, select the option **Statistics**. In Screen 4.16, select the following options:

   Percentiles Values: Quartiles
   Central Tendency: Mean, median, mode
   Dispersion: Std. deviation, variance, range, minimum, maximum

   Click **Continue**.

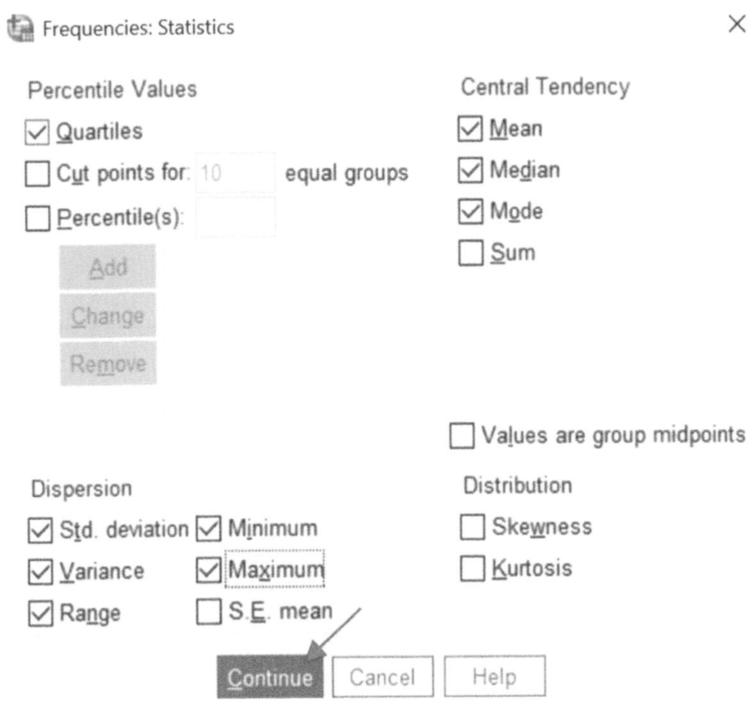

Screen 4.16 Frequencies – Statistics

4. Under the main dialogue box of **Frequencies**, click **Charts** to select the type of chart to be generated (see Screen 4.17).

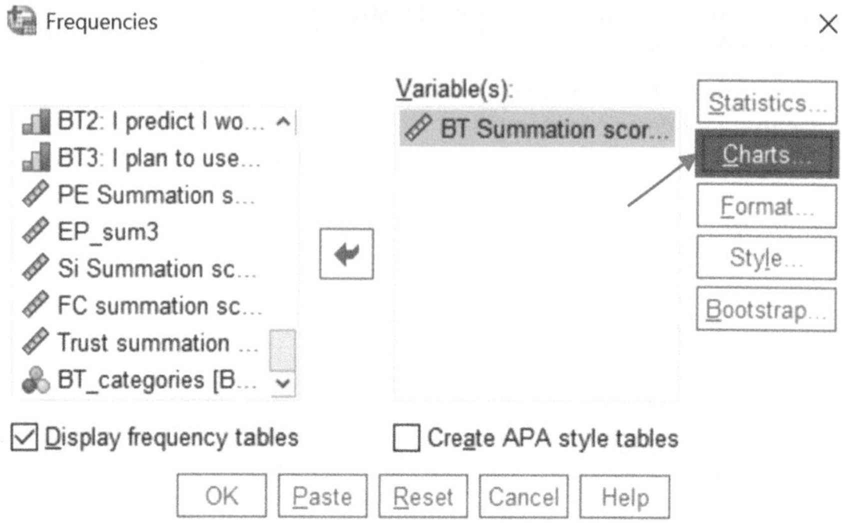

Screen 4.17 Frequencies – Charts

5. Under the **Frequencies: Charts**, select **Histograms**, and check the box Show normal curve on the histogram (see Screen 4.18). Then, click **Continue** to go back to the main dialogue box.

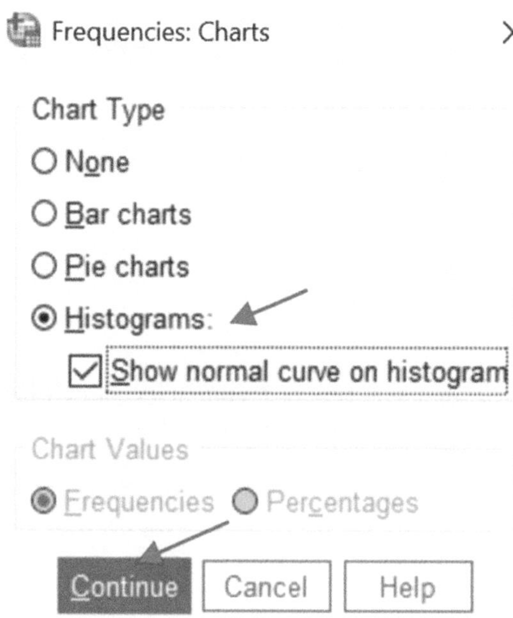

Screen 4.18 Frequencies – Charts continue

6. Click **OK** to perform the test.

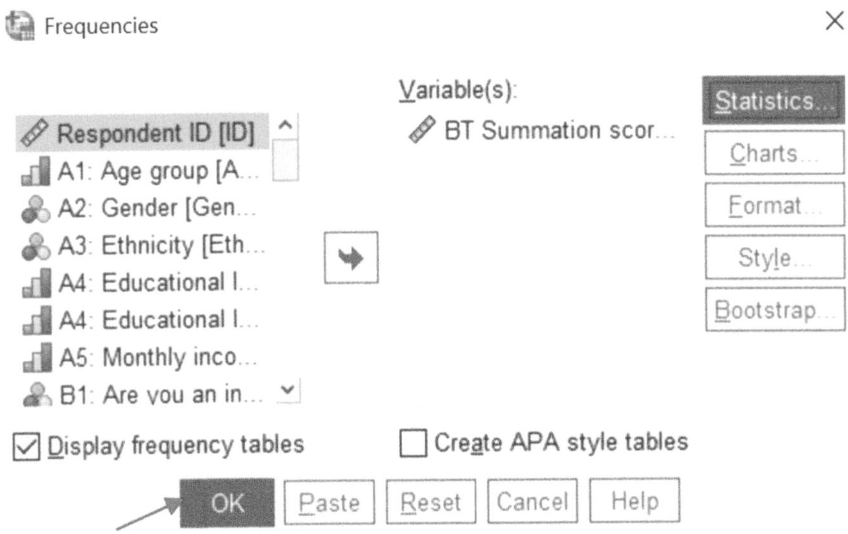

Screen 4.19 Run the frequency analysis

4.3.1 *Generated SPSS outputs*

The outputs of the frequency analysis as shown in Screen 4.20 to Screen 4.22 (see pages 58–59).

4.3.2 *Sample of write-up in APA format*

This section demonstrates report writing in APA style and presents a table of descriptive statistics and also a histogram with a normal curve.

Based on the 5-point Likert scale of behavioural intention score, the minimum score was 3 and the maximum of 15, and this gives a range of 12 (see Table 4.3). The median behavioural intention score was 11.00 with a standard deviation of 2.81. The standard deviation of 2 means that about 95% of the data falls within 2 standard deviations of the mean. The mean behavioural intention score was 10.07 shows that, overall, the behavioural intention score is acceptably high.

The 25th percentile of the behavioural intention score is 8, and the 75th percentile is 12, and thus, the inter-quartile range (IQR) is 4. The values obtained for the 25th and 75th percentiles indicated that 50% of respondents have behavioural intention scores between 8 and 12. From the histogram

BT Summation score: BT1 to BT3

N	Valid	200
	Missing	0
Mean		10.07
Median		11.00
Mode		12
Std. Deviation		2.811
Variance		7.900
Range		12
Minimum		3
Maximum		15
Percentiles	25	8.00
	50	11.00
	75	12.00

Screen 4.20 Statistics

BT Summation score: BT1 to BT3

		Frequency	Percent	Valid Percent	Cumulative Percent
Valid	3	4	2.0	2.0	2.0
	5	4	2.0	2.0	4.0
	6	32	16.0	16.0	20.0
	7	3	1.5	1.5	21.5
	8	9	4.5	4.5	26.0
	9	26	13.0	13.0	39.0
	10	17	8.5	8.5	47.5
	11	16	8.0	8.0	55.5
	12	64	32.0	32.0	87.5
	13	11	5.5	5.5	93.0
	14	5	2.5	2.5	95.5
	15	9	4.5	4.5	100.0
	Total	200	100.0	100.0	

Screen 4.21 Frequency and percentage

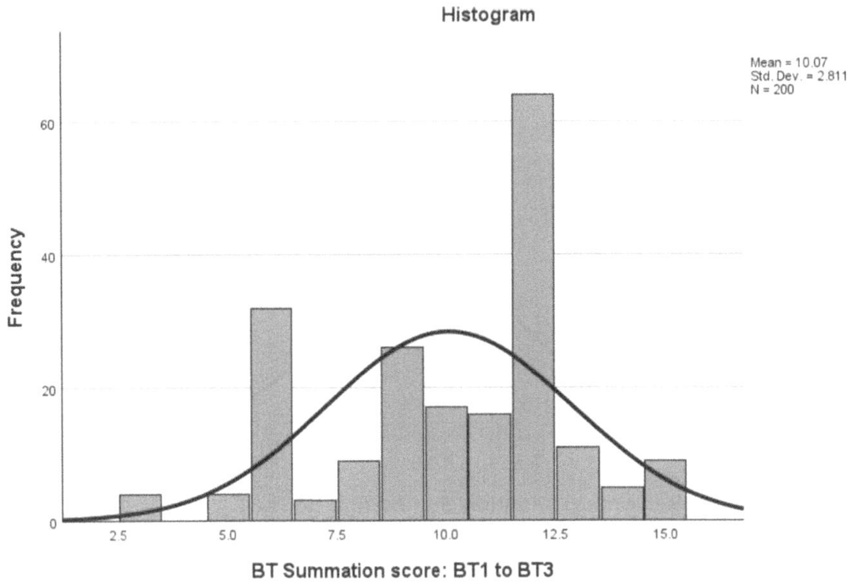

Screen 4.22 Histogram with a normal curve

Table 4.3 Descriptive statistics of the behavioural summation score and the selected variables

Descriptive statistics	*Behavioural intention score*	*Variable 2*	*Variable 3*
Mean	10.07		
Median	11.00		
Standard Deviation	2.81		
Minimum	3		
Maximum	15		
Interquartile Range (IQR)	4		
Percentile			
25th	8		
50th	11		
75th	12		

illustration, the behavioural intention score is considered as normally distributed and negatively skewed, as depicted in Figure 4.1.

4.4 Descriptive statistics and Z-score

The earlier section of this chapter provides the detailed steps and a sample write-up on how to generate descriptive statistics of both measures of central tendency and variability. Generating a Z-score is an additional function of the

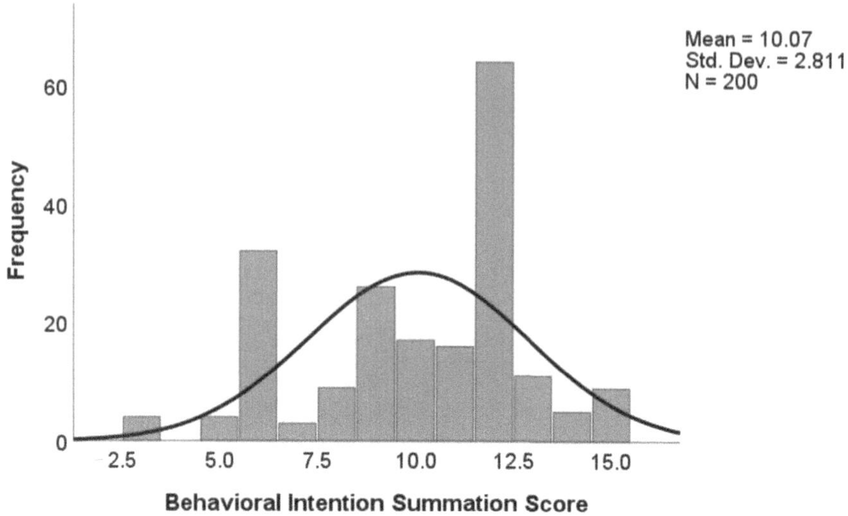

Figure 4.1 Histogram of behavioural intention score with a normal curve

descriptive statistics. Z-score, the standardized values, can be saved as a new variable for further analysis, especially in multiple regression. The Z-score is useful to compare samples from different populations and to identify outliers during data screening. The acceptable range of Z-score is between −3 and +3; any cases more than +3 and less than −3 will be identified as an outlier.

SPSS steps using the Internet Banking data file.

1. On the **Analyze** menu, go to **Descriptive Statistics**, and then select **Descriptives** (see Screen 4.23).

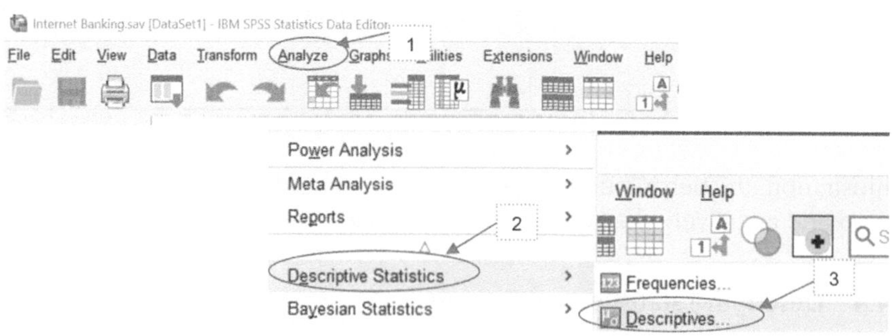

Screen 4.23 Analyze – open descriptive statistics

The **Descriptives** dialogue box appears (Screen 4.24).

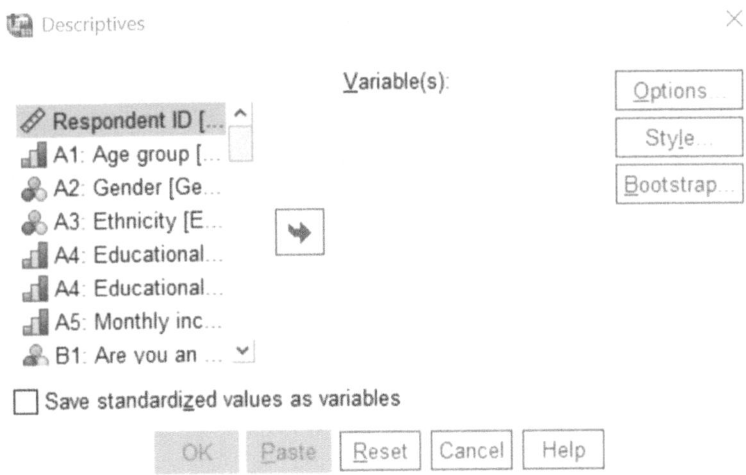

Screen 4.24 Descriptives

2. From the variable list box, select and move the variable **BT Summation** into the **Variable(s)**. Tick the box **Save standardized values as variables** (see Screen 4.25).

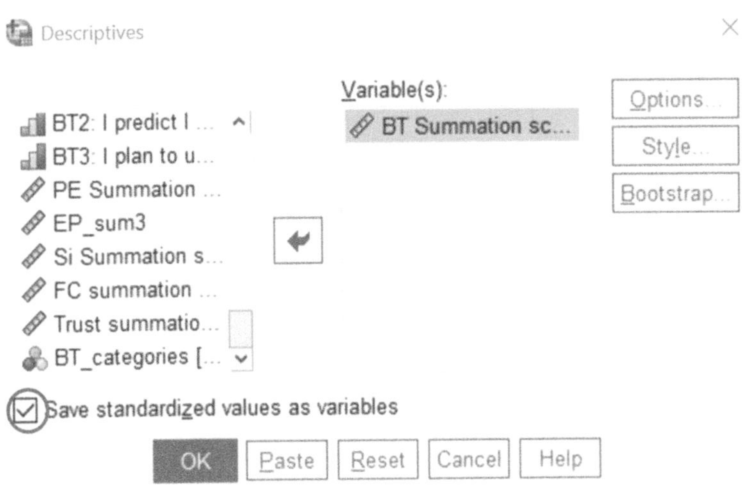

Screen 4.25 Descriptives – save standardized values as variables

3. Next, click the **Options** button (see Screen 4.26). In Screen 4.27 **Descriptives: Options**, the **Mean**, **Std. Deviation**, **Minimum**, and **Maximum** are selected by default. Thus, just click **Continue**.

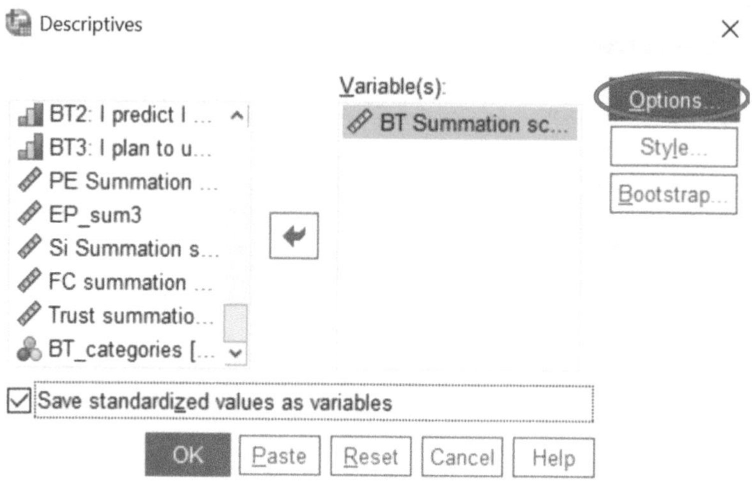

Screen 4.26 Descriptives – Options

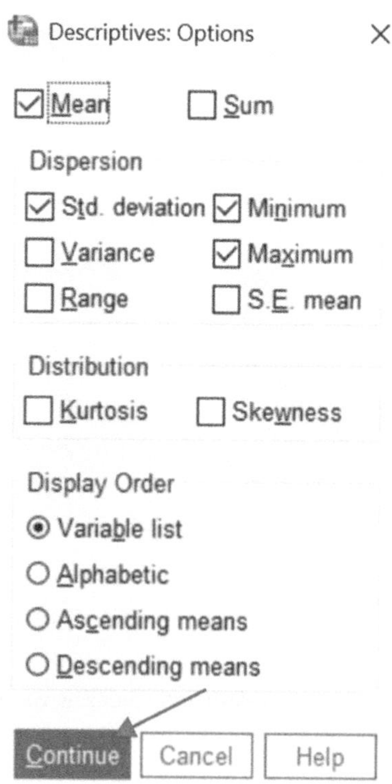

Screen 4.27 Descriptives – Select options

4. Click **OK** to run the analysis.

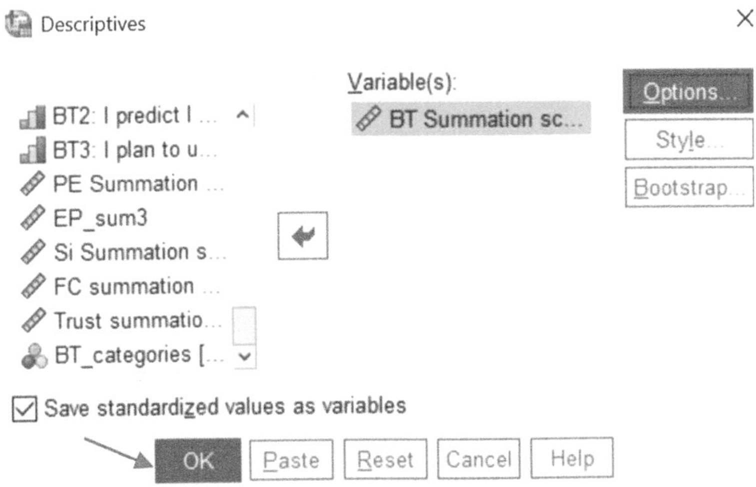

Screen 4.28 Execute the analysis

4.4.1 SPSS outputs

From the SPSS outputs, there is a summary of descriptive statistics that includes the total valid cases, minimum and maximum values, as well as the mean and standard deviation of the behavioural summation score (see Screen 4.29).

	N	Minimum	Maximum	Mean	Std. Deviation
BT Summation score: BT1 to BT3	200	3	15	10.07	2.811
Valid N (listwise)	200				

Screen 4.29 Descriptive Statistics

On the variable view

Then, let's check the data file and look for variable view and data view. An additional row has been inserted at last (Screen 4.30). It is a new variable, the Z-score for the behavioural intention summation score.

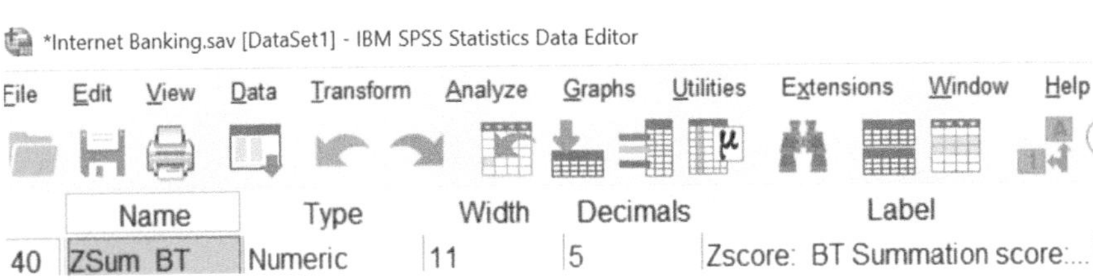

Screen 4.30 Variable view

On the data view

An additional column has been inserted; at last, it is valued for the Z-score (Screen 4.31).

Upon checking, the Z-score for the behavioural intention summation score falls between −2.51 and 1.76. To check the Z-score range, only right-click the Z-score to sort ascending or descending. Alternatively, run a quick analysis that is **Descriptives . . .** under **Descriptives Statistics** under the **Analyze** toolbar. Since the Z-score is within the acceptable range of between −3 and +3, it has sufficient evidence to claim that there is no outlier for the behavioural intention summation score.

ZSum_BT
-1.09046
-.02313
.68843
-.37890
-.37890
.68843
1.04421
1.39999
1.04421

Screen 4.31 Variable view

5 All about T-tests

5.1 One-sample T-test

Learning outcomes

At the end of this section, participants will be able to:

5.1.1 Know the purposes and requirements of the one-sample T-test,
5.1.2 Demonstrate the one-sample T-test using SPSS,
5.1.3 Interpret the generated outputs, and
5.1.4 Report the interpreted results in APA format.

5.1.1 Purpose and requirements of the one-sample T-test

The one-sample T-test is used to determine the statistical mean difference of a hypothesised or known population mean. In this test, the selected variable is also known as a test variable, and it is used to compare with a "test value" (i.e. hypothesised or known population mean).

Since this is a parametric test, one of the key assumptions is that the data follow a normal distribution. The test variable is continuous data that is either an interval or a ratio scale. All observations have value and should not contain outliers.

The effect size (Cohen's d) is the mean difference between two groups, which is expressed in standard deviation units. Based on the guideline of Cohen (1988), a Cohen's d value close to 0.2 is a small effect, close to 0.5 is a moderate effect, and close to 0.8 is considered a large effect. Any values larger than 1 are considered a very large effect.

Working example

A student explores the shopping behaviour of shoppers and wants to know if the unknown mean income is different from $1500. The following is the research objective:

Research objective:

To identify whether the mean income of the shoppers is <u>different</u> from $1500.

DOI: 10.4324/9781003673903-5

Test variable: income
Test value: $1500

To run this test, we only use **one** selected test variable. Before that, refer to the EDA analysis to check the normality of the test variable.

5.1.2 Steps to run a one-sample T-test

1. On the menu bar, select **File**, **Open,** and **Data** (see Screen 5.1).

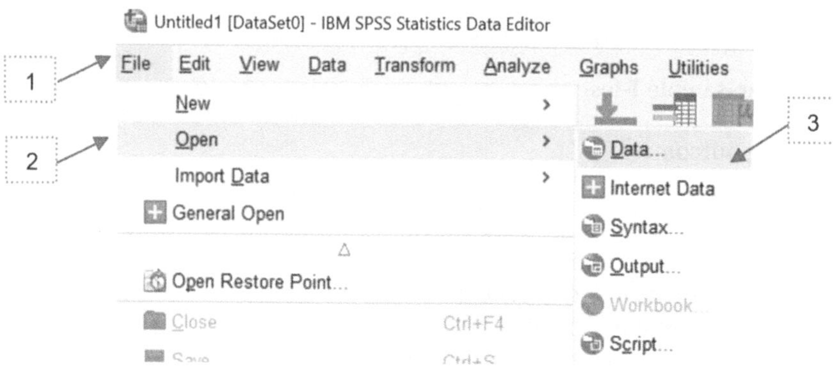

Screen 5.1 Data editor

2. In the sample data SPSS folder, select the **Shopping behaviour.sav** to open the data file (see Screen 5.2).

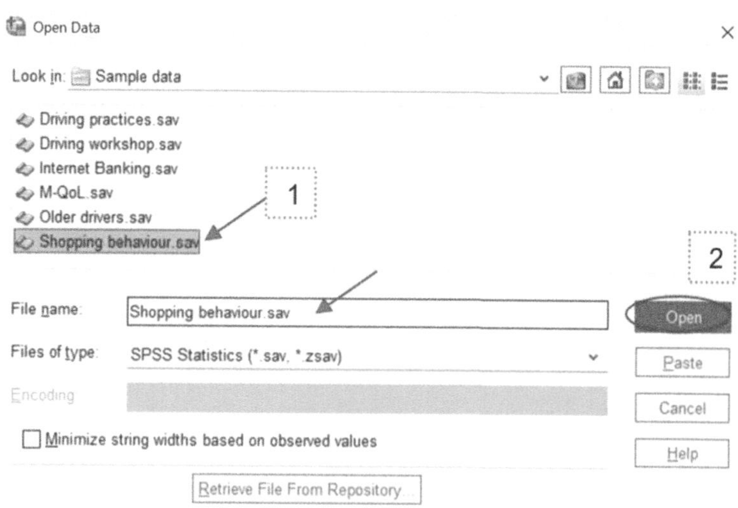

Screen 5.2 Open data

3. Go to the main menu, click **Analyze**, go to **Compare Means and Proportions**, and choose the second option, **One-Sample T Test** (see Screen 5.3).

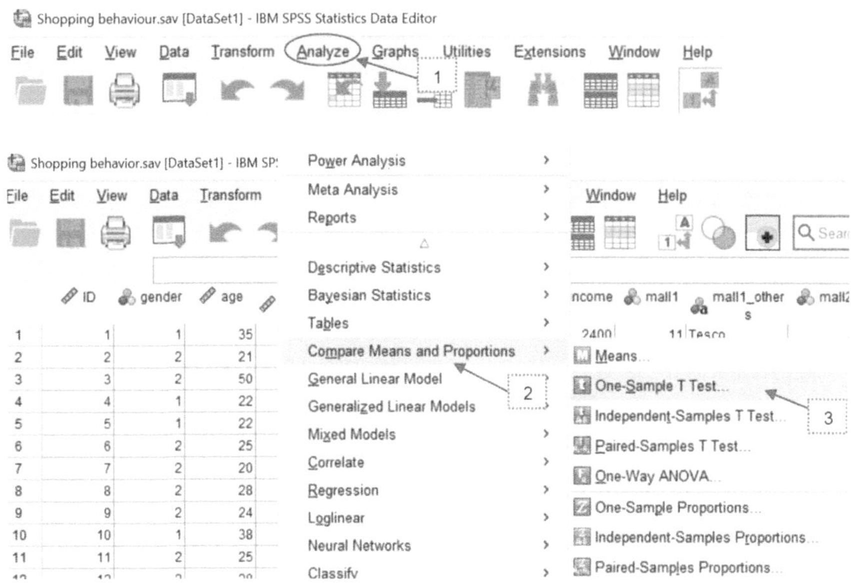

Screen 5.3 Select analysis

Screen 5.4 will appear.

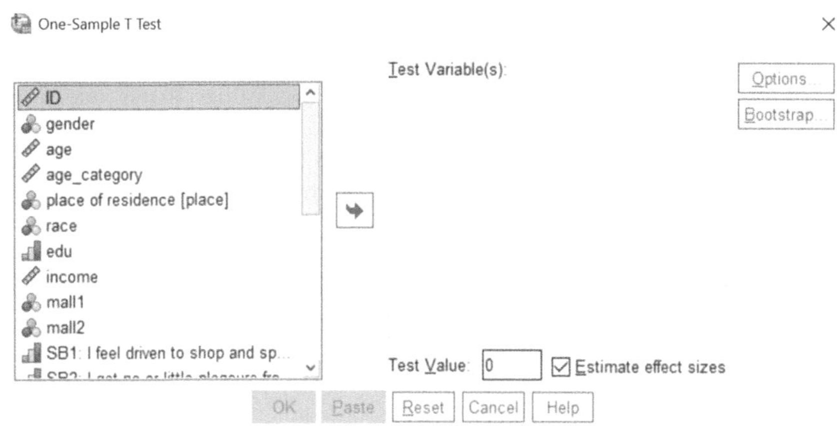

Screen 5.4 One-sample T-test

4. Find the **income** variable in the variable list box, and transfer it into the **Test Variable(s)**. Then, type 1500 in the **Test Value** box (see Screen 5.5). Lastly, click **OK** to run the test.

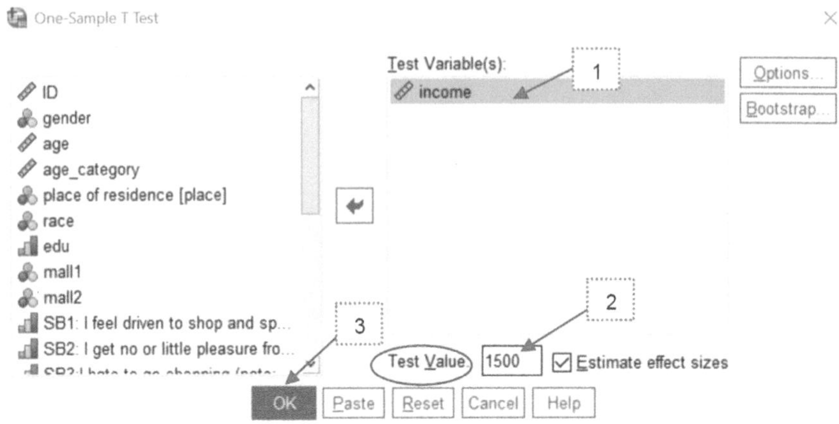

Screen 5.5 One-sample T-test – insert test variable and test value

5.1.3 Hypothesis testing and interpretation of one-sample T-test SPSS outputs

1. State the hypotheses based on the research objective in Chapter 5.1.1.

   Null hypothesis
   $\mathbf{H_0}$: The mean income is equal to $1500.
   Alternative hypothesis
   $\mathbf{H_a}$: The mean income is not equal to $1500.

2. Based on the SPSS output (see Screen 5.6), decide on the null hypothesis by using a 0.05 level of significance and conclude.

   The one-sample T-test provides two significance values, that is, one-sided p and two-sided p. When the hypothesis did not indicate the test value's trend to be higher or lower, compared to the 0.05 level of significance to the two-sided p-value. Alternatively, if the hypothesis has specified the test value to be higher or lower, then refer to the one-sided p-value. In this example, a two-sided p-value will be reported.

   The obtained p-value ($p = .0001$) was smaller than 0.05 (α). The obtained t-statistic was larger than 1.96 ($t = 4.810$). Therefore, the null hypothesis (**$\mathbf{H_0}$) is rejected**, and thus, H_a is supported.

   We can conclude that the mean income of shoppers is different from or not equal to $1500. The actual mean income is $2234.46 ($SD = 1313.63$), which is very much higher than the test value.

3. Since there is a significant difference between the test variable and the test value. We can further check the Cohen's d value to know the effect size of the mean difference.

One-Sample Statistics

	N	Mean	Std. Deviation	Std. Error Mean
income	74	2234.46	1313.625	152.706

One-Sample Test

	Test Value = 1500						
			Significance		Mean	95% Confidence Interval of the Difference	
	t	df	One-Sided p	Two-Sided p	Difference	Lower	Upper
income	4.810	73	<.001	<.001	734.459	430.12	1038.80

One-Sample Effect Sizes

		Standardizer[a]	Point Estimate	95% Confidence Interval Lower	Upper
income	Cohen's d	1313.625	.559	.312	.803
	Hedges' correction	1327.317	.553	.309	.794

Screen 5.6 Outputs of the one-sample T-test

Compare this obtained eta-squared value, Cohen's d = <u>0.56</u>, to the guideline by Cohen (1988). The obtained effect size is considered to be <u>moderate</u>. This implies that the mean difference between the mean income and the population mean income is <u>moderate</u>.

5.1.4 Report one-sample T-tests in APA format

Before conducting the one-sample T-test, preliminary checking was done to check the normality of the data set to ensure there is no violation. A one-sample T-test was used to examine whether the mean income of the shoppers is different from \$1500. The shoppers reported higher income (M = \$2234.46, SD = 1313.63) than the population mean income, $t(73) = 4.81$, $p < 0.05$. Cohen's d (d = 0.56) implies that the strength of the difference between the mean income and the population mean income is moderate. The positive value of the eta-squared value reflects that the population's mean income is higher than the sample mean income of shoppers. Generally, the shoppers have a relatively higher income.

5.2 Independent-samples T-test

Learning outcomes

At the end of this section, participants will be able to:

5.2.1 Recognise the purposes of the independent-samples T-test,
5.2.2 Demonstrate the independent-samples T-test step by step using SPSS,
5.2.3 Interpret the various outputs, and
5.2.4 Write an example write-up in APA format.

5.2.1. Purpose and requirements of the independent-samples T-test

An independent-samples T-test analysis will be done to identify if there are any significant differences in the means for two groups in the variable of interest.

Two assumptions of the independent-samples T-test are:

1. Normal distribution
2. Equality of variance

Requirements of the variables' scale measurement

Independent variable: Interval/ratio (continuous data)
Dependent variables: Nominal/ordinal with two groups only (categorical data with two categories)

Examples of research objectives:

1. To identify the differences between gender and shopping behaviour
2. To identify the differences between internet banking adopters and non-adopters in terms of trust

Refer to chapter EDA, the normality test of gender and shopping behaviour was performed. The results show that the data have met the assumption of normality.

Thus, the independent-samples T-test is used to test research objective no. 1.

5.2.2 Steps to run an independent-samples T-test analysis

1. On the menu bar, select **File**, **Open,** and **Data** (see Screen 5.7).

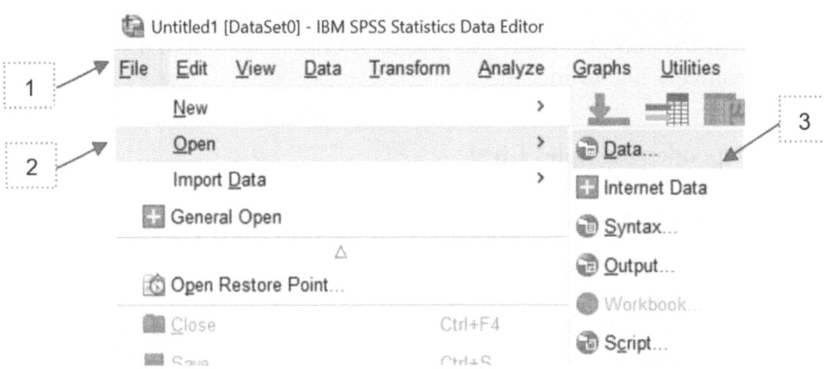

Screen 5.7 Data editor

2. The **Open Data** dialogue box appears (see Screen 5.8). In the sample data SPSS folder, select the **Shopping behaviour** data file. To open the file, click **Open**.

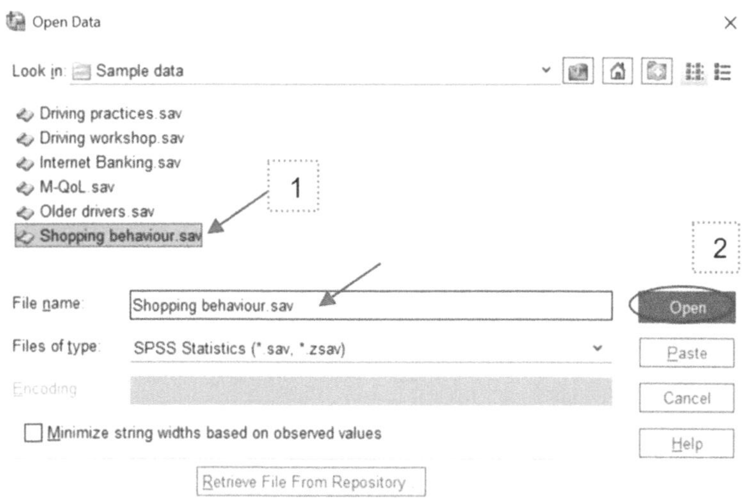

Screen 5.8 Open data

3. Go to the main menu, choose **Analyze**, then **Compare Means**, and follow with **Independent-Samples T Test** (see Screen 5.9).

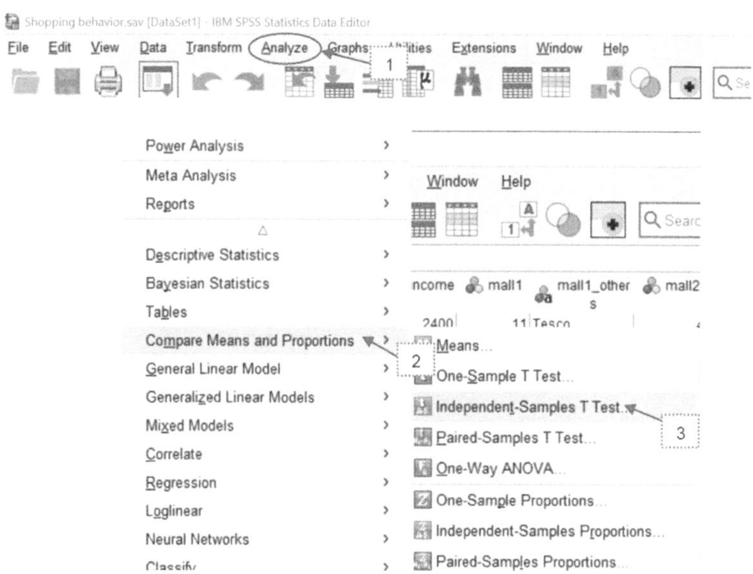

Screen 5.9 Main Menu: Choose Independent-Samples T Test

4. Select the test variable **(SB_sumscore)** from the variable list box. Move it into **Test Variable(s)** (see Screen 5.10).

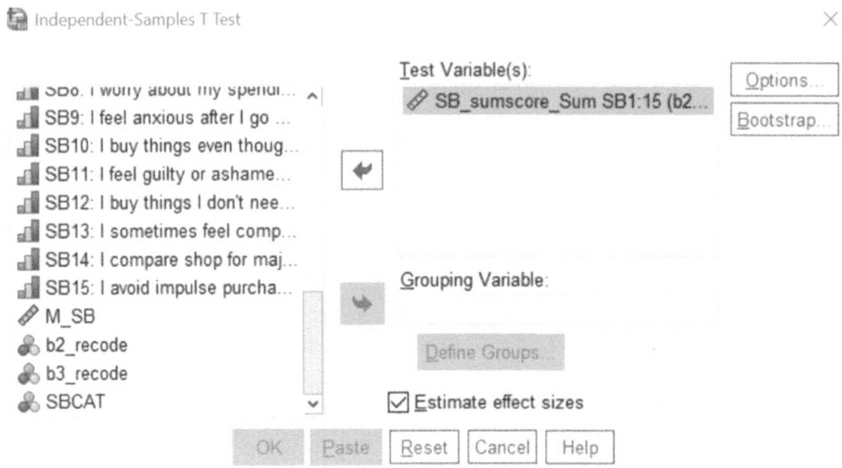

Screen 5.10 Independent-samples T-test

5. Next, select the test variable (**gender**) and move it into the **Grouping Variable** (see Screen 5.11).

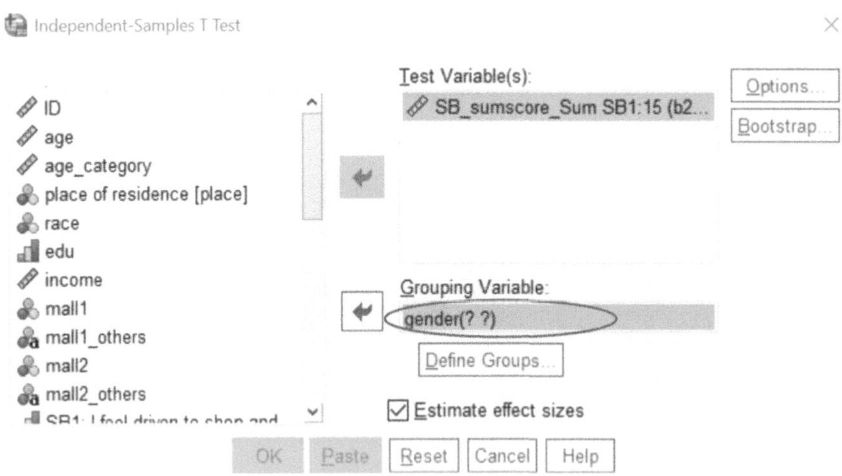

Screen 5.11 Grouping variable

6. Click the **Define Groups** option under the grouping variable (see Screen 5.12).

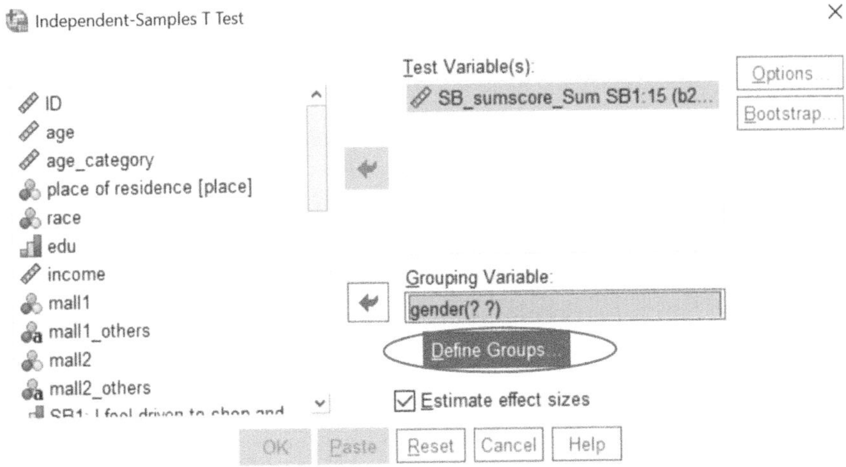

Screen 5.12 Define Groups

7. In older SPSS versions, the user needs to key in specified values for **Group 1 and 2** based on the original data coding. In SPSS version 30, the system will automatically enter the predetermined value for each group. In this case, see Screen 5.13, the data coding of gender has been defined as **(1, 2)**. The value should be consistent with the data coding. Let's say the data coding for gender is 0 and 1, regardless of the value, either for male or female. So, remember to amend the values for both groups to **(0, 1)**.

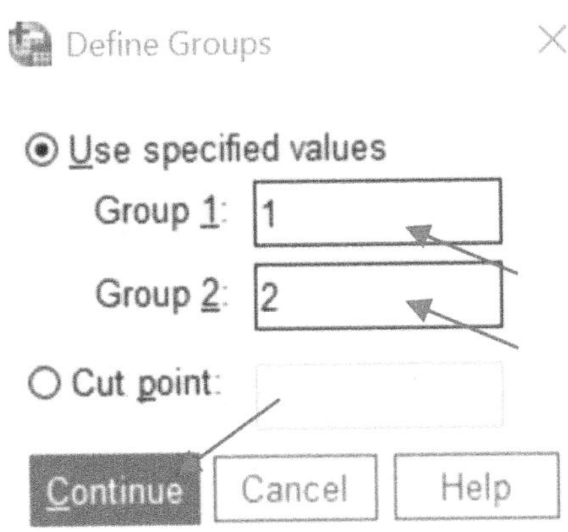

Screen 5.13 Define Groups

Then, click **Continue** to go back to the main **Independent-Sample T Test** dialogue box.

8. To run the test, click the **OK** option in Screen 5.14.

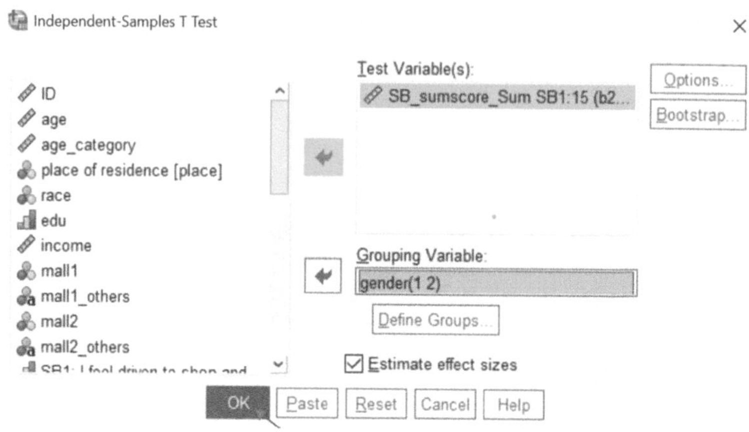

Screen 5.14 Run the test

5.2.3 Interpretation of independent-samples T-test SPSS outputs

1. State the null and alternative hypotheses based on research objective 1.

   Null hypothesis

   H_o: The mean shopping behaviour score for female and male shoppers is equal.

or

   H_o: There is no mean difference in the shopping behaviour score between female and male shoppers.

   Alternative hypothesis

   H_a: The mean shopping behaviour score for female and male shoppers is not equal.

or

   H_a: There is a significant mean difference in shopping behaviour scores between female and male shoppers.

2. Refer to Screen 5.15, determine which *t*-value should be reported, and justify it.

   Refer to **equal variance not assumed** (not equal variance); the *t*-value of –3.029 shall be reported.

Group Statistics

	gender	N	Mean	Std. Deviation	Std. Error Mean
SB_sumscore_Sum SB1:15 (b2,b3 recode)	Male	36	50.67	4.296	.716
	Female	38	54.55	6.562	1.065

A B

Independent Samples Test

		Levene's Test for Equality of Variances		t-test for Equality of Means							
						Significance				95% Confidence Interval of the Difference	
		F	Sig.	t	df	One-Sided p	Two-Sided p	Mean Difference	Std. Error Difference	Lower	Upper
SB_sumscore_Sum SB1:15 (b2,b3 recode)	Equal variances assumed	8.240	.005	-2.996	72	.002	.004	-3.886	1.297	-6.472	-1.300
	Equal variances not assumed			-3.029	64.167	.002	.004	-3.886	1.283	-6.449	-1.323

H_o → Equal variances assumed

H_a → Equal variances not assumed

Screen 5.15 SPSS outputs

The results of Levene's test for equality of variance indicate that the mean variances of shopping behaviour score for both groups of shoppers are not equal ($F = 8.24$, $p = .005$).

(Note: Refer to **A** [Levene's test for equality of variance in Screen 5.15]; the same decision rule of hypothesis testing applies. Since $p = 0.005$ is smaller than the 0.05 level of significance, we reject the null hypothesis, H_o, and accept H_a: equal variance not assumed. Thus, report the second row of the T-test of Equality of Means in **B** in Screen 5.15.)

3. Decide on the null hypothesis and justify.

> *How to determine which significance p to be reported?*
>
> The hypothesis aims to check whether the mean shopping behaviour is equal or not. There is no specific trend to test which gender higher or lower shopping behaviour score. Thus, a two-sided p-value will be reported compared 0.05 level of significance. If the hypothesis wants to test a higher or lower score between these two groups (a directional outcome), then refer to the one-sided p-value.

The null hypothesis (**H_o) is rejected**, **so H_a is supported.**

The obtained t-statistic was larger than 1.96 ($t = -3.029$). The obtained p-value ($p = .004$) was smaller than 0.05 (α).

4. Provide the general conclusion based on the null hypothesis decision.

Since H_a is supported, a significant mean difference exists between the shopping behaviours score of male and female shoppers.

In fact, female shoppers (54.55) scored higher in shopping behaviour than male shoppers (50.67).

5. Since there is a significant difference between the selected variables, it is important to further discuss the effect size by Cohen's d (see Screen 5.16).

Compare this obtained Cohen's d = – 0.70 to the guideline by Cohen (1988). The obtained effect size is considered to be close to large. The negative value of Cohen's d means that the second group mean, which is female, is higher. This implies that the mean difference between male and female shoppers in terms of shopping behaviour score is considered large.

		Standardizer[a]	Point Estimate	95% Confidence Interval Lower	95% Confidence Interval Upper
SB_sumscore_Sum SB1:15 (b2,b3 recode)	Cohen's d	5.577	-.697	-1.164	-.225
	Hedges' correction	5.636	-.689	-1.152	-.222
	Glass's delta	6.562	-.592	-1.064	-.113

Screen 5.16 Independent samples effect sizes

6. Refer to Screen 5.17, the output of a normality test. Would you consider this analysis valid?

		Kolmogorov-Smirnov[a]			Shapiro-Wilk		
	gender	Statistic	df	Sig.	Statistic	df	Sig.
Sum SB1:15 (b2,b3 recode)	Male	.133	36	.110	.960	36	.214
	Female	.127	38	.126	.943	38	.052

Screen 5.17 Tests of normality

Analysis validity:

Yes, this independent-samples T-test analysis is valid because the data fulfilled the normality assumption. (Refer to Chapter 1, EDA analysis, to recap how to run a normality test.)

5.2.4 *Present the complete write-up in APA format*

An independent-samples T-test was performed to identify the mean differences of shopping behaviour scores among male and female shoppers. Before the analysis, tests of normality and equality of variance were conducted to check if the data fulfilled the assumptions of the analysis. The results indicate that no serious violations were noted. A quick inspection of the boxplots displayed in Figure 5.1 shows that the mean shopping behaviour score for female shoppers is higher than that of male shoppers. There was a significant

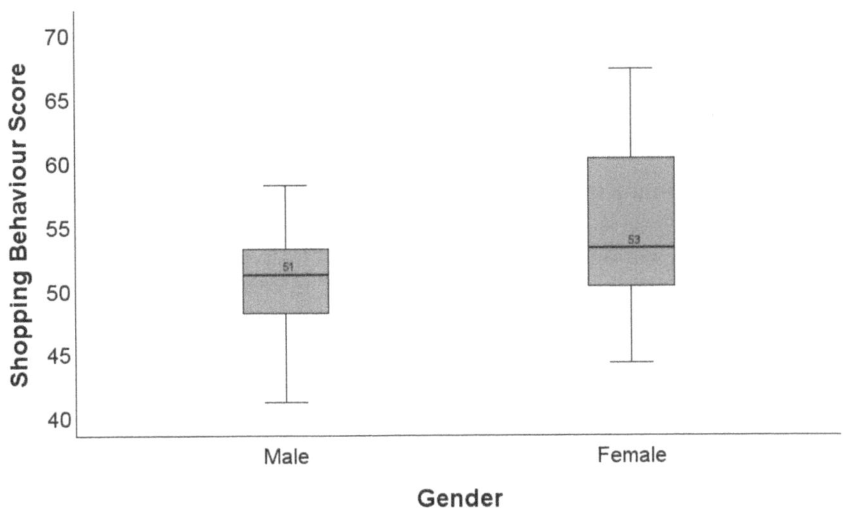

Figure 5.1 The boxplot of mean shopping behaviour scores by gender

difference in the mean of shopping behaviours score for male shoppers (M = 50.67, SD = 4.30) and female shoppers [M = 54.55, SD = 6.56; t (72) = −3.03, $p < .05$]. In comparing these two means, the study found that female shoppers were more addicted to shopping than male shoppers. Additionally, the magnitude of the mean differences among genders was large (d = 0.70) based on the guideline by Cohen (1988).

Note: If the boxes have similar heights, the variances across the groups are considered equal (homogeneity). Conversely, if the variance across groups is not homogeneous, the boxes have different heights. Thus, it is not a surprise when Levene's test is significant.

Present a summary table of the independent-samples T-test in APA format

Table 5.1 Results of the T-test between gender and shopping behaviour

Gender	*n*	*M*	*SD*	*t*	*p*
Male					
Female					

*$p < .05$, significant at the 0.05 level of significance (2-tailed)

5.3 Paired-samples T-test

Learning outcomes

At the end of this section, participants will be able to:

5.3.1 Recognise the purposes and requirements of the paired-samples T-test,
5.3.2 Demonstrate how to run a paired-samples T-test using SPSS,
5.3.3 Interpret the outputs from the test, and
5.3.4 Report paired-samples T-test in APA format.

5.3.1 Purpose and requirements of paired-samples T-test

This test can examine the mean difference of two variables with a continuous scale (i.e. interval or ratio). Frequently, we want to know the mean difference, typically before and after (or pre- and post-) of those who undergo a certain treatment, medication, workshop, training, etc.

Assumptions:

- The cases/observations are independent of each other.
- The variables should meet the normal distribution.
- No existence of outliers.

Tips for a successful paired-samples test

It must contain a complete set of values to run the test. You need to make sure there is a value for pre- and post- (or before and after) in all participants or

Table 5.2 Example of research objectives that required a paired-samples test

Research objective examples	*Independent variable*	*Dependent variable*
1. To identify the mean difference in driving knowledge of drivers before and after attending the driving workshop.	Pre-driving knowledge score/pre-workshop	Post driving knowledge score/post-workshop
2. To determine the mean difference of BMI index before and after the weight management programme.	Pre-BMI index	Post-BMI index
3. To examine the mean difference of glucose level before and after medication treatment.	The glucose level before	The glucose level after

experiments. (The pairwise deletion method is recommended to handle the missing data in these selected variables.)

The scenario of a working example

Based on the example of research objective 1, let's assume you have conducted a driving workshop to improve the driving knowledge of older drivers (output). Thus, two times data collection was conducted, and the driving knowledge before and after the driving workshop was assessed.

5.3.2 *Steps to run a paired-samples T-test using the driving workshop data set*

1. On the menu bar, select **File**, **Open,** and **Data** (see Screen 5.18).

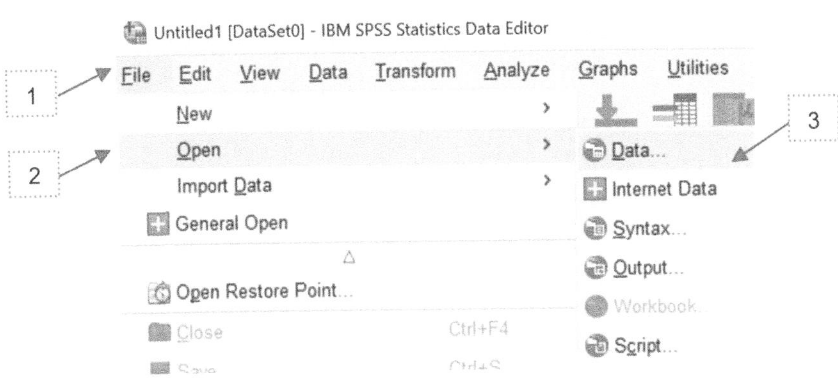

Screen 5.18 Data editor

2. In the folder sample data, select the **Driving Workshop** data file and click **Open** to open the data file (see Screen 5.19).

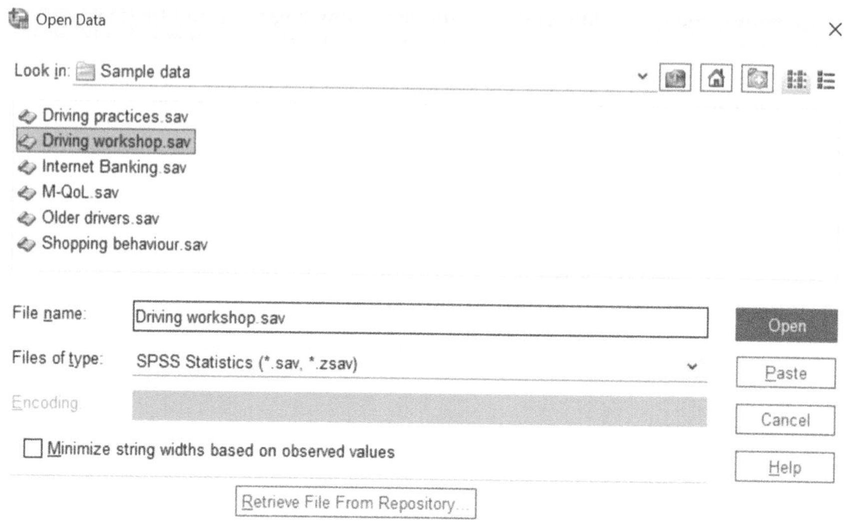

Screen 5.19 Open data

3. Go to the main menu, click **Analyze,** then go to **Compare Means**, and click **Paired-Samples T Test** (see Screens 5.20–5.21).

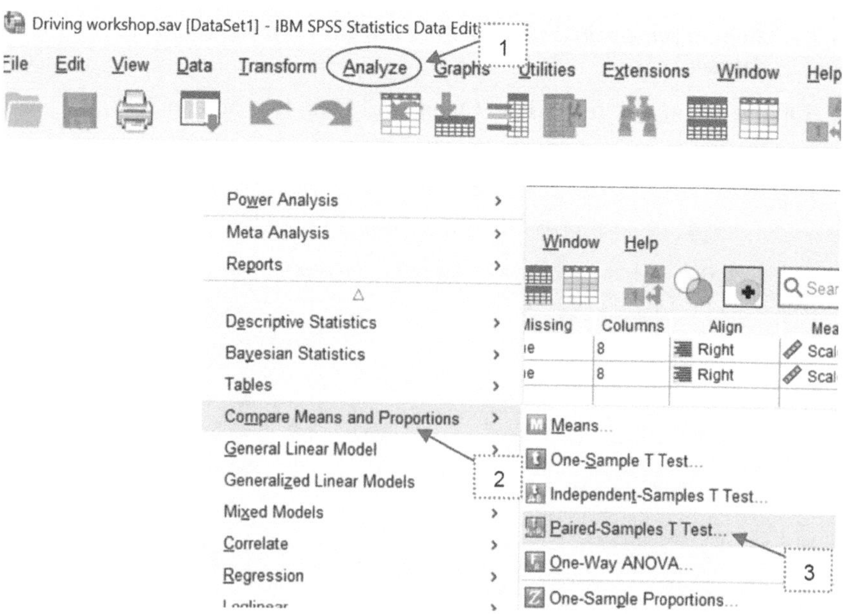

Screen 5.20 Select Paired-Samples T Test

The **Paired-Samples T test** dialogue box appears (see Screen 5.21).

Screen 5.21 Paired-Samples T Test dialogue box

4. From the variable list box, move the variable "**Before**" into Variable 1, and the variable "**After**" into Variable 2 under **Paired Variables** (see Screen 5.22).

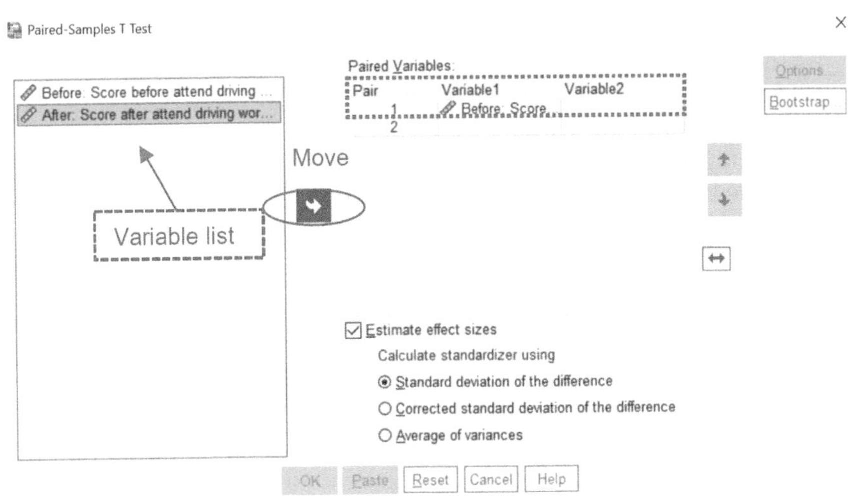

Screen 5.22 Enter variables

5. Click **OK** to run the test (see Screen 5.23).

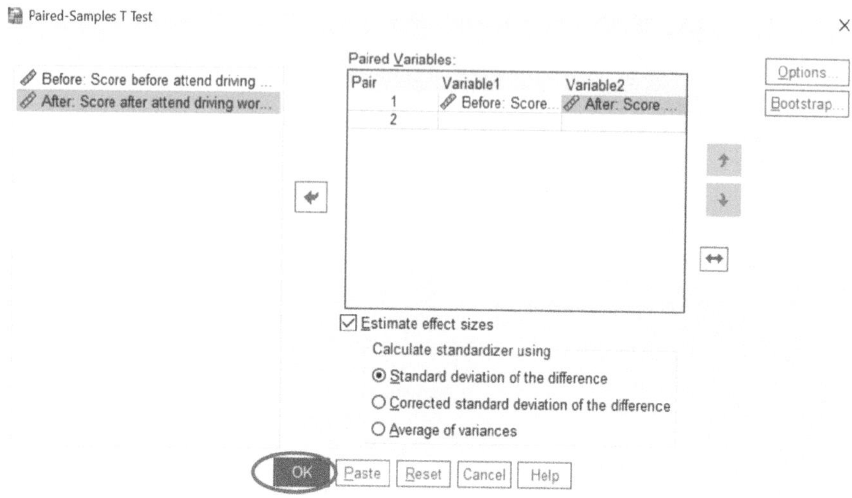

Screen 5.23 Run the analysis

5.3.3 Hypothesis testing and interpretation

1. State the alternative hypotheses for the test.

   Null hypothesis (H_0):

   The mean for the pre-driving knowledge score is equal to the pre-driving knowledge score.

   or

   There is no difference in the mean of the pre-driving knowledge score and the post-driving knowledge score.

   Alternate hypothesis (H_a):

   The mean for the pre-driving knowledge score is not equal to the post-driving knowledge score.

   or

   There is a significant difference in the mean of the pre-driving knowledge score and the post-driving knowledge score.

2. Based on the SPSS outputs of the **Paired Samples Test** (see Screen 5.24), decide on your hypothesis and justify.

> *How to determine which significance p to be reported*
>
> Since the hypothesis has no specific trend to test whether the mean of pre-driving knowledge score is higher or lower. Thus, a two-sided p-value will be reported with compared 0.05 level of significance. If the hypothesis wants to test a higher or lower score between pre- and post-, then refer to the one-sided p-value.

Paired Samples Statistics

		Mean	N	Std. Deviation	Std. Error Mean
Pair 1	Before: Score before attend driving workshop	9.50	30	2.713	.495
	After: Score after attend driving workshop	14.37	30	2.266	.414

Paired Samples Correlations

		N	Correlation	Significance One-Sided p	Significance Two-Sided p
Pair 1	Before: Score before attend driving workshop & After: Score after attend driving workshop	30	.631	<.001	<.001

df (degree of freedom) =N-1

Paired Samples Test

		Paired Differences Mean	Std. Deviation	Std. Error Mean	95% Confidence Interval of the Difference Lower	Upper	t	df	Significance One-Sided p	Two-Sided p
Pair 1	Before: Score before attend driving workshop - After: Score after attend driving workshop	-4.867	2.177	.398	-5.680	-4.054	-12.243	29	<.001	<.001

Screen 5.24 Outputs of paired-samples T-test

The null hypothesis (H_o) is rejected, and hence, the alternative hypothesis (H_a) is supported. The |t-statistic| was larger than 1.96 ($t = -12.243$) and the observed p-value ($p = .0001$) is smaller than the alpha value at 0.05. Thus, the obtained mean difference of −4.867 was considered large.

3. Conclude the analysis.

   Since H_a is supported, the conclusion is that there is a significant difference in the mean of the pre-driving knowledge score and the post-driving knowledge score.

   An inspection of the mean scores revealed that the mean for post-driving knowledge score ($M = 14.37$) is significantly higher than the pre-driving knowledge score ($M = 9.50$). This implies that the driving workshop has successfully improved the driving knowledge of the participants.

4. To know the effect size of the mean difference, refer to the point estimate by Cohen's d in Screen 5.25.

From the SPSS output, the Cohen's d of −2.24, which is larger than 1, indicates a substantial difference between pre- and post-groups (refer to Chapter 5.1.1). In this case, the effect size is very large. This implies that the magnitude of the mean difference for driving knowledge score between the pre- and post-workshop was very large. The negative value of Cohen's d means that the second mean post-workshop is higher. Given a large effect size, the implemented driving workshop was very effective in improving the driving knowledge of the participants.

			Standardizer[a]	Point Estimate	95% Confidence Interval Lower	Upper
Pair 1	Before: Score before attend driving workshop - After: Score after attend driving workshop	Cohen's d	2.177	-2.235	-2.904	-1.554
		Hedges' correction	2.236	-2.177	-2.829	-1.514

Screen 5.25 Paired-samples effect sizes

5.3.4 Sample write-up for the paired-samples T-test in APA format

The first research objective 1 in this study is to identify the mean difference in driving knowledge of drivers before and after attending the driving workshop. A paired-samples T-test was conducted to compare the mean differences of driving knowledge to achieve the first research objective. See Figure 5.2; the boxplot illustrates that the mean driving knowledge post-workshop is much higher than pre-workshop. An exploratory descriptive analysis was conducted to examine the normality assumption. There is no violation of the normal distribution, or the data distribution is normal. An inspection of the mean scores revealed that the mean for the post-driving knowledge score (M = 14.37; SD = 2.71) is significantly higher than the

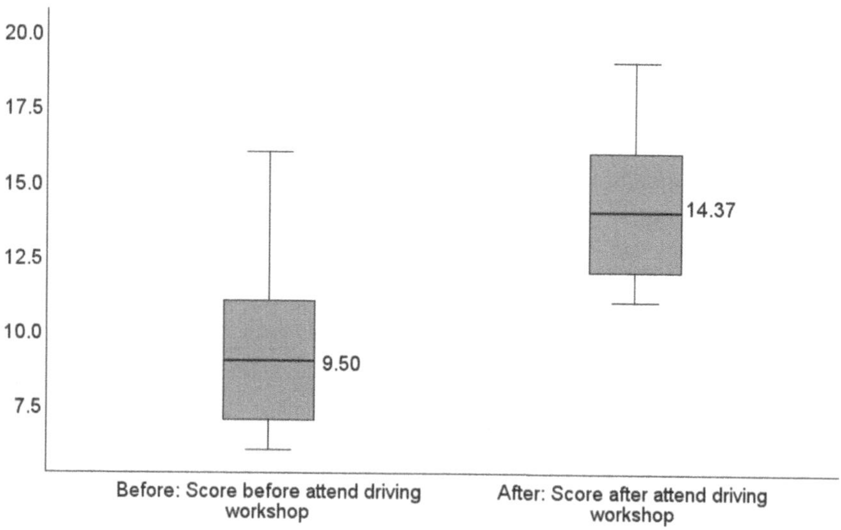

Figure 5.2 The boxplot of the mean driving knowledge before and after attending the driving workshop

pre-driving knowledge score (*M* = 9.50; *SD* = 2.23). This implies that the driving workshop has successfully improved the driving knowledge of the participants. There was a significant difference in driving knowledge in the pre-workshop (*M* = 14.37; *SD* = 2.71) and post-workshop [*M* = 9.50; *SD* = 2.23; $t\,(29) = -12.243$, $p < .05$]. A further inspection of the two means suggests that the mean for driving knowledge before the workshop was significantly higher than that of those who had attended the driving workshop. The magnitude of the mean difference for driving knowledge score between the pre- and post-workshop was large ($d = -2.24$). This implies that the implemented driving workshop was very effective in improving the driving knowledge of the participants.

Steps to create the comparative boxplot

1. Click **Analyze**, go to **Descriptive Statistics**, and choose **Explore** (see Screen 5.26).

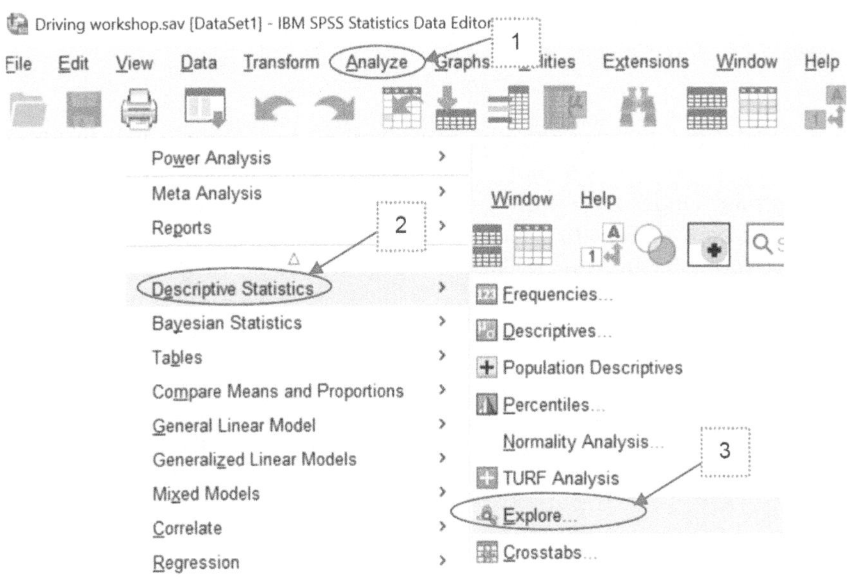

Screen 5.26 Select the Explore function

2. Move the variables before and after into the **Dependents List** (see Screen 5.27).

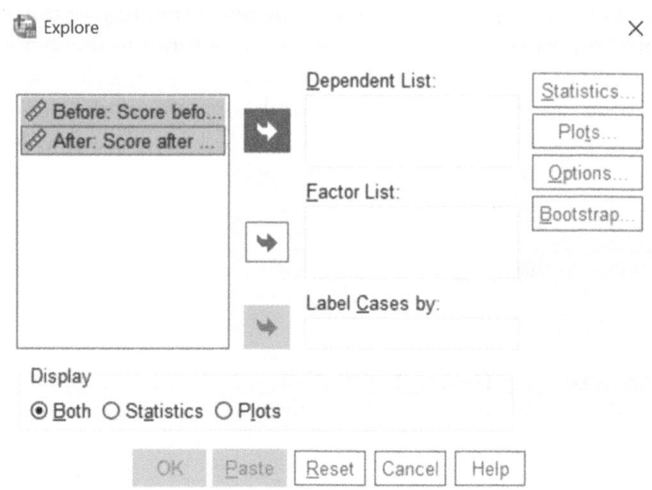

Screen 5.27 Move variables into Explore

3. To make the two variables display on the same chart. Click the **Plots** button, under the **Boxplots**, and change the selection to **Dependents Together**. Under **Descriptive**, uncheck **Stem-and-leaf** and click **Continue** (see Screens 5.28–5.29).

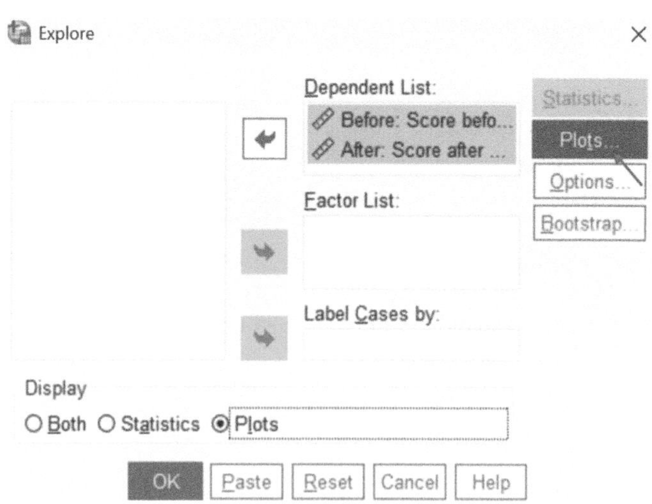

Screen 5.28 Explore: Plots

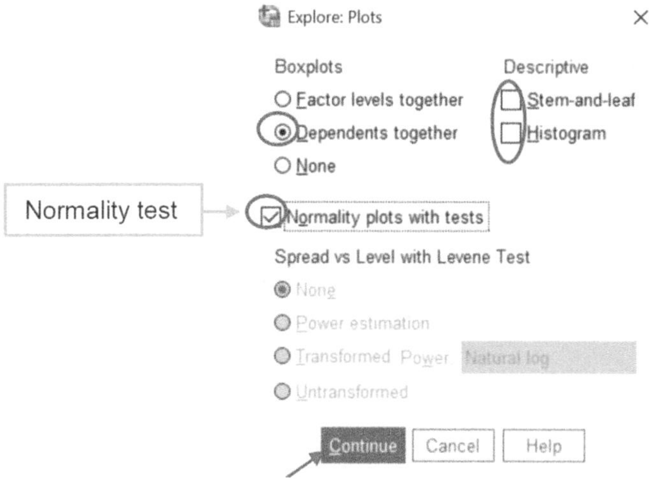

Screen 5.29 Continue the option of Plots

4. Click **OK** to generate the comparative boxplot (see Screen 5.30).

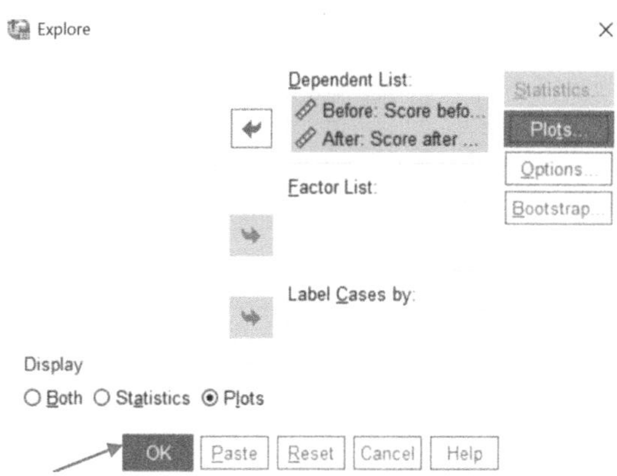

Screen 5.30 Generate the boxplot

6 One-way ANOVA

Learning outcomes

At the end of this section, the participants will be able to:

6.1 Recognise the purposes and requirements of one-way ANOVA analysis,
6.2 Demonstrate one-way ANOVA analysis using SPSS,
6.3 Understand the SPSS outputs of the one-way ANOVA analysis, and
6.4 Report the analysis in APA format.

6.1 Purpose and requirements of one-way ANOVA analysis

The purpose of this analysis is to test the significant mean differences for more than two groups on a dependent variable with an interval or ratio-scaled variable.

There are **two assumptions** to be fulfilled: normality test and equality of variance. In other words, the data should be normally distributed and have equal variance.

The following is an example of a research objective that required a one-way ANOVA analysis:

To identify the mean difference between age groups and behavioural intention to use internet banking among Kuala Lumpur residents.
Independent variable:

Age groups (grouping of ages is expected, such as 21–30, 31–40, 41–50 years)
Dependent variable:
Behavioural intention

** Behavioural intention was collected using multiple questions with 5-point Likert scales. Transformation (Transform-Compute) is needed for the behavioural intention summation score.*

DOI: 10.4324/9781003673903-6

Tips for a successful one-way ANOVA analysis

Data transformation is needed for ungrouped and uncategorized continuous data, multiple items scale, etc. The following are the frequently used functions under transform.

1. Recode into a different variable using the range option, such as variables for income level, age group, and stress level.
2. The option of Compute is needed when:

   - A variable measured by the scale/multiple questions (must have the same response options) sums up the responses for each question to get a summation score. If there is a negative statement, use the recoded response.
   - A set of relevant questions was asked, or information compiled during data collection. For example, the height and weight of the respondents were collected during data collection, and thus, the researcher wants to calculate the BMI index.

6.2 Demonstrate one-way ANOVA analysis using SPSS

The working data is from Yeoh and Chan (2011) about behavioural intention towards internet banking.

1. On the menu bar, select **File**, **Open**, and **Data** (see Screen 6.1).

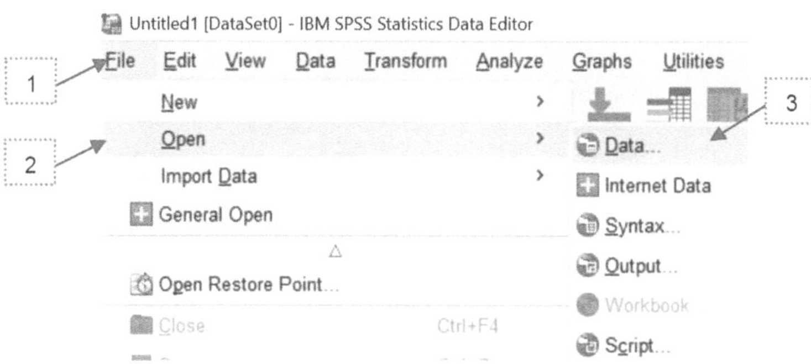

Screen 6.1 Data editor

2. In the sample data SPSS folder, select the **Internet Banking** data file and then click **Open** to open up the data file (see Screen 6.2).

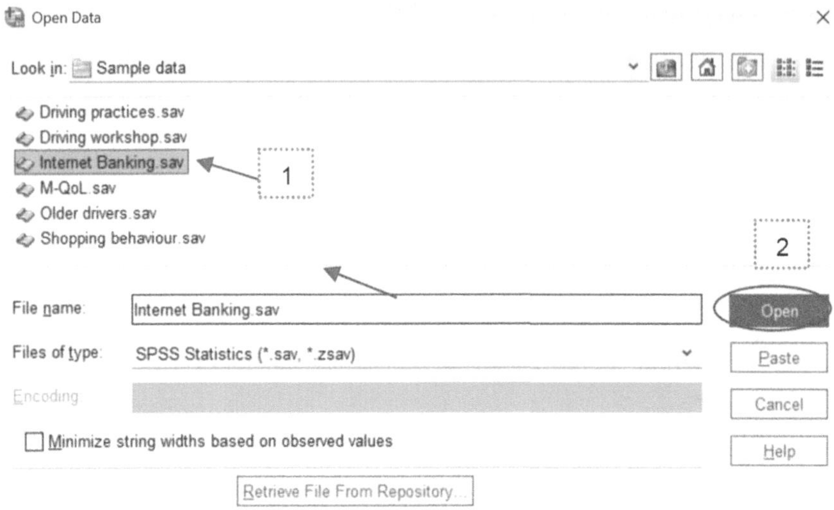

Screen 6.2 Open data

3. On the **Analyze** menu, choose **Compare Means,** and then click **One-Way ANOVA** (see Screen 6.3).

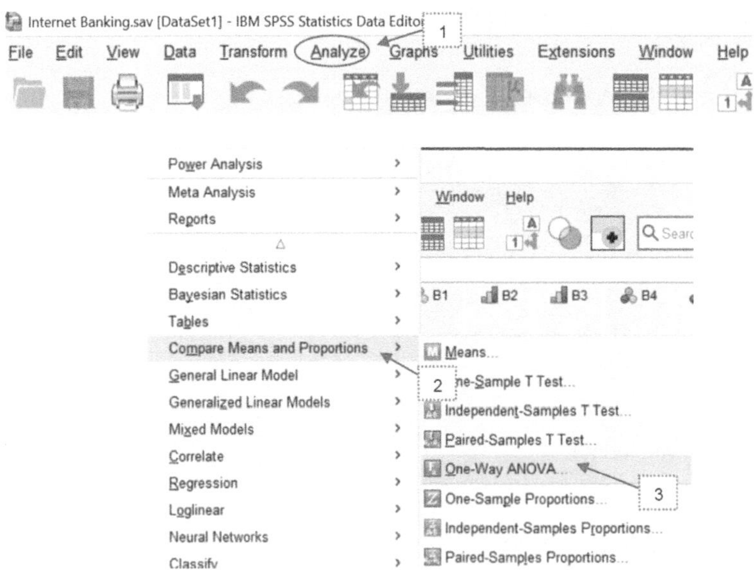

Screen 6.3 Select analysis

The dialogue box **One-Way ANOVA** will appear (see Screen 6.4).

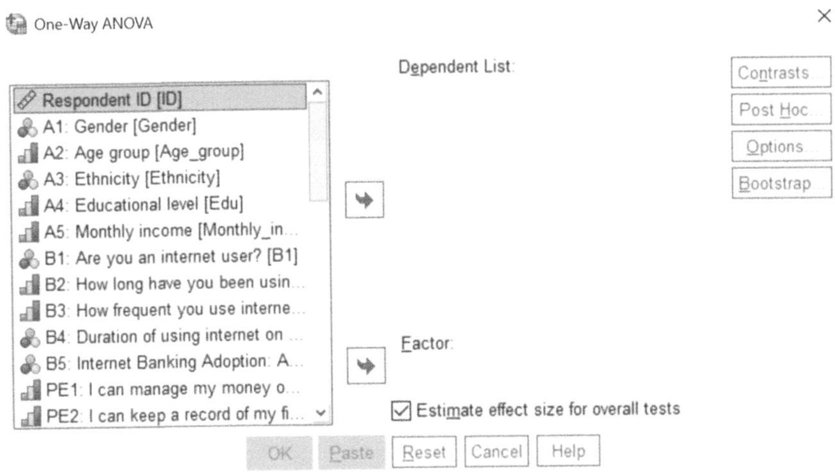

Screen 6.4 One-way ANOVA

4. Search the **BT Summation score** (dependent variable) in the variable list box and move it into the **Dependent List** box (see Screen 6.5).

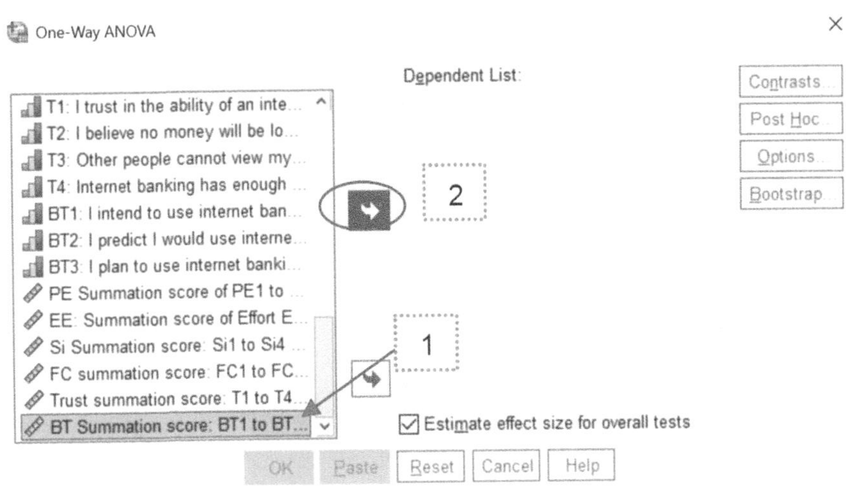

Screen 6.5 Move variable to Dependent List

5. Search **Age group** (factor/independent variable) in the variable list box. Move it into the **Factor** box (see Screen 6.6).

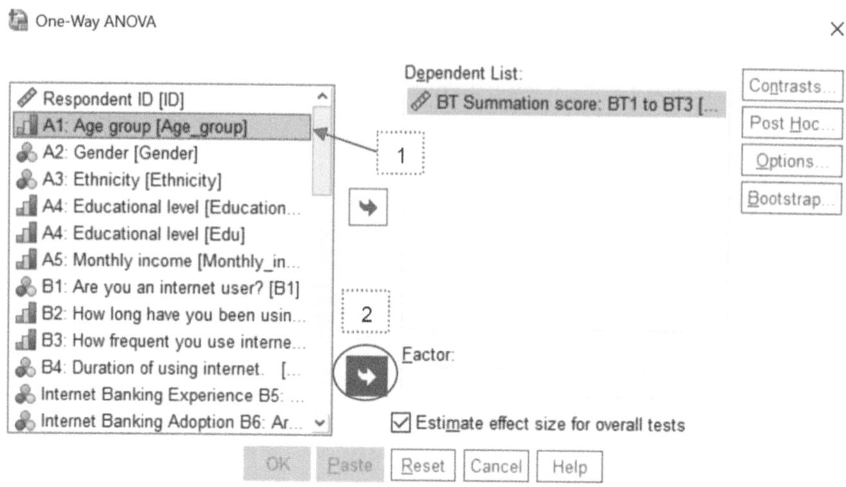

Screen 6.6 Move variable to the Dependent List

6. To generate the results of significant pair comparisons. Click the button **Post Hoc** to open its dialogue box (see Screen 6.7).

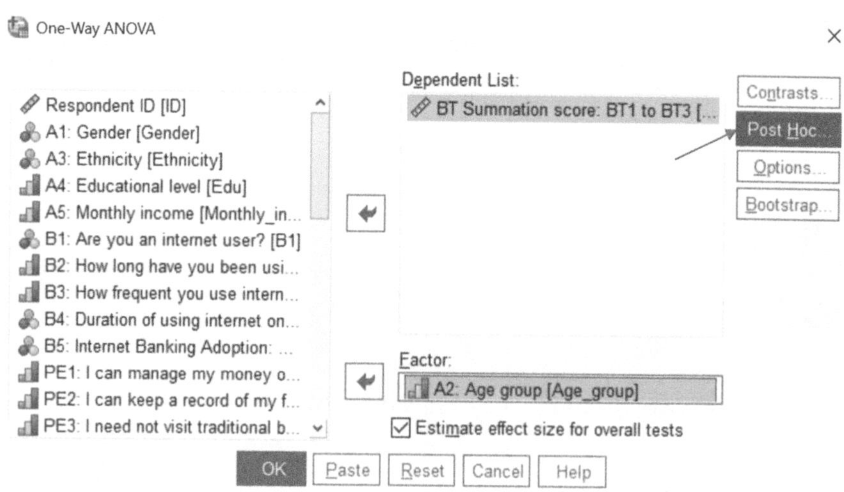

Screen 6.7 Select Post Hoc

7. Under **Equal Variances Assumed**, check the options of **Bonferroni** and **Scheffe**. Under the **Null Hypothesis test**, select the significance level for the post hoc test at the 0.05 level (see Screen 6.8). Then, click **Continue** to return to the main dialogue box.

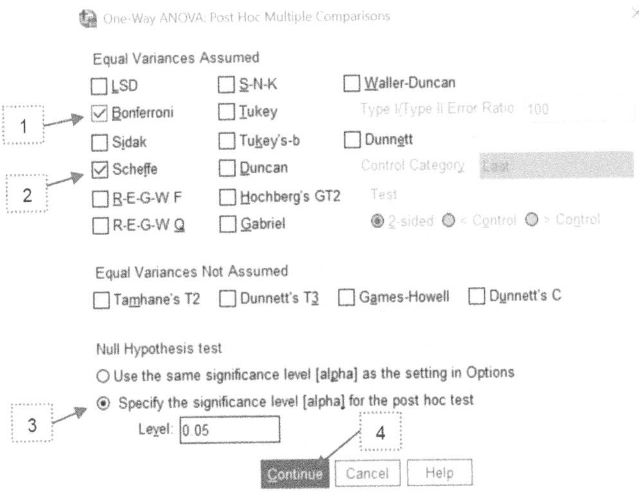

Screen 6.8 Post hoc multiple comparisons

8. In the main dialogue box, click **Options** to view the details (see Screen 6.9).

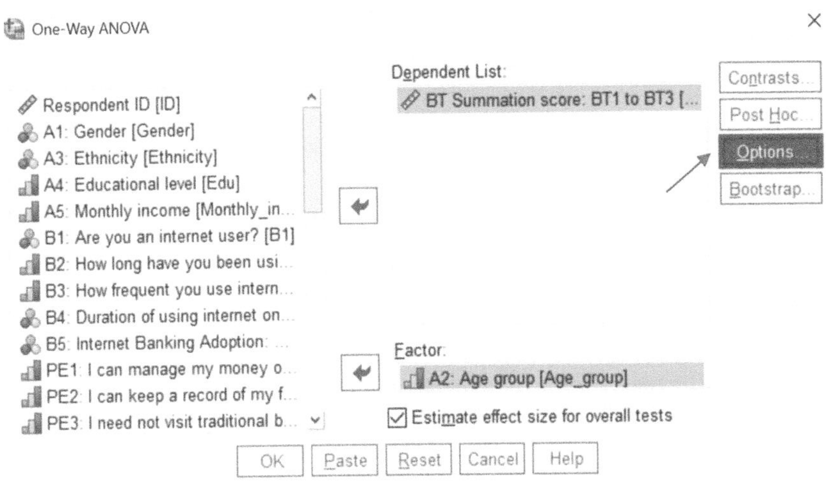

Screen 6.9 Options

9. Under the options of **Statistics** (see Screen 6.10), check the box of **Descriptive** and **Homogeneity of variance test** (a.k.a. Levene's test). Check the **Means plot** box, and click **Continue** to go back to the main dialogue box.

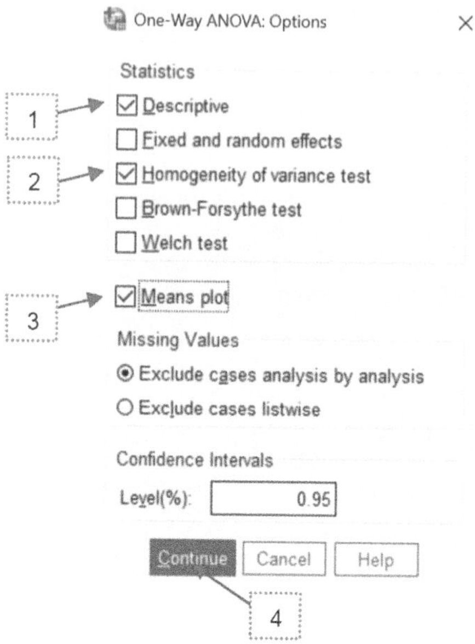

Screen 6.10 Choose the options

10. Click **OK** to generate results for one-way ANOVA (see Screen 6.11).

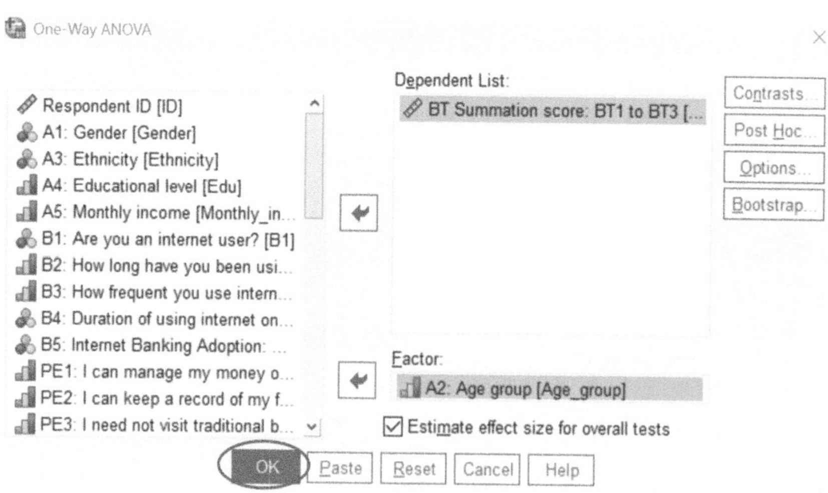

Screen 6.11 Run one-way ANOVA analysis

6.3 Understand the SPSS outputs of the one-way ANOVA analysis

The researchers are interested in knowing the differences between age groups and their behavioural intentions **(BT Summation score)** to use internet banking.

Screen 6.12 is part of the SPSS outputs in the one-way ANOVA test.

Descriptives

BT Summation score: BT1 to BT3

	N	Mean	Std. Deviation	Std. Error	95% Confidence Interval for Mean Lower Bound	95% Confidence Interval for Mean Upper Bound	Minimum	Maximum
21-30	89	11.00	2.321	.246	10.51	11.49	5	15
31-40	62	9.40	2.972	.377	8.65	10.16	3	15
41-50	49	9.20	2.951	.422	8.36	10.05	3	15
Total	200	10.07	2.811	.199	9.67	10.46	3	15

ANOVA

BT Summation score: BT1 to BT3

	Sum of Squares	df	Mean Square	F	Sig.
Between Groups	141.276	2	70.638	9.725	<.001
Within Groups	1430.879	197	7.263		
Total	1572.155	199			

Screen 6.12 Descriptives and ANOVA

Research objective: To identify the mean difference between age groups and behavioural intention to use internet banking among Kuala Lumpur residents.

1. Provide the null and alternative hypotheses based on the research objective.

   Null hypothesis (H_0):

   There is no significant mean difference in behavioural intention among respondents across different age groups.

   or

   The mean of behavioural intention for respondents aged 21 to 30 years, 31 to 40 years, and 41 to 50 years is the same or equal.

   Alternative hypothesis (H_a):

   There is a significant mean difference in behavioural intention among respondents across different age groups.

   or

   Not all means are equal.
2. Decide on the null hypothesis and provide a conclusion.

   The null hypothesis (H_0) is rejected, and hence, H_a is supported.

Justification: The F-statistic/ratio was considered as large ($F = 9.725$) as compared to the cut-off of 3.96, and the obtained p-value ($p < .0001$) was smaller than 0.05 (α).

Conclusion

Since the H_a is supported, the means are unequal.

Note: *Further inspection on the significant differences in the groups is needed when the null hypothesis (*H_o*) is rejected; otherwise, further inspection is not needed.*

3. Based on the outputs of multiple comparisons in Screen 6.13, identify the pairs of means that are significantly different from each other. Provide the relevant write-up and justification.

Dependent Variable: BT Summation score: BT1 to BT3

	(I) A2: Age group	(J) A2: Age group	Mean Difference (I-J)	Std. Error	Sig.	95% Confidence Interval Lower Bound	Upper Bound
Scheffe	21-30	31-40	1.597*	.446	.002	.50	2.70
		41-50	1.796*	.479	.001	.61	2.98
	31-40	21-30	-1.597*	.446	.002	-2.70	-.50
		41-50	.199	.515	.928	-1.07	1.47
	41-50	21-30	-1.796*	.479	.001	-2.98	-.61
		31-40	-.199	.515	.928	-1.47	1.07
Bonferroni	21-30	31-40	1.597*	.446	.001	.52	2.67
		41-50	1.796*	.479	<.001	.64	2.95
	31-40	21-30	-1.597*	.446	.001	-2.67	-.52
		41-50	.199	.515	1.000	-1.04	1.44
	41-50	21-30	-1.796*	.479	<.001	-2.95	-.64
		31-40	-.199	.515	1.000	-1.44	1.04

*. The mean difference is significant at the 0.05 level.

Screen 6.13 Multiple comparisons

Checking on the results of Scheffie's post hoc multiple comparison tests, the means of behavioural intention for the following pairs were found to be statistically significantly different:

1. 21 to 30 years old ($M = 11.00$, $SD = 2.32$) and 31 to 40 years old ($M = 9.40$, $SD = 2.97$), because the obtained p-value is smaller ($p = .002$) than 0.05 (α). Therefore, the mean difference ($MD = 1.60$) is considered large.
2. 21 to 30 years old ($M = 11.00$, $SD = 2.32$) and 41 to 50 years old ($M = 9.20$, $SD = 2.95$), because the obtained p-value is smaller ($p = .001$) than 0.05 (α). Therefore, the mean difference ($MD = 1.80$) is considered large.

Checking on the results of Bonferroni post hoc multiple comparison tests, the mean difference of behavioural intention for the following pairs was found to be insignificant:

3. 31 to 40 years old (M = 15.46, SD = 21.65) and 41 to 50 years old (M = 21.29, SD = 28.15), the obtained p-value is very much larger (p = 1.000) than 0.05 (α). Thus, the mean difference is considered small (MD = 0.199).

   Since there is a significant mean difference between groups. Refer to Screen 6.14 for the ANOVA effect size. What is the effect size of the mean difference between age groups and behavioural intention?

ANOVA Effect Sizes[a]

		Point Estimate	95% Confidence Interval Lower	95% Confidence Interval Upper
BT Summation score: BT1 to BT3	Eta-squared	.090	.024	.166
	Epsilon-squared	.081	.014	.158
	Omega-squared Fixed-effect	.080	.014	.157
	Omega-squared Random-effect	.042	.007	.085

a. Eta-squared and Epsilon-squared are estimated based on the fixed-effect model.

Screen 6.14 ANOVA effect sizes

Eta-squared (η^2) cutoffs (Cohen, 1988)

- $\eta^2 < 0.01$: "negligible"
- $0.01 \leq \eta^2 < 0.06$: "small"
- $0.06 \leq \eta^2 < 0.14$: "medium"
- $\eta^2 \geq 0.14$: "large"

The obtained eta-squared value (η^2 = 0.09) is considered to be moderate. Therefore, a big/large mean difference existed between the selected variables.

4. Recall the assumptions of one-way ANOVA, which are the test of normality and equality of variance (i.e. homogeneity of variance). To generate the test of normality, normal Q-Q plots, and box plots, refer to Chapter 1 to explore the data under descriptive analysis.

Refer to Screens 6.15–6.21 for EDA analysis, and identify if this one-way ANOVA analysis is valid or appropriate. What is your suggestion to solve the problem, if there is any?

	A1: Age group	Kolmogorov-Smirnov[a]			Shapiro-Wilk		
		Statistic	df	Sig.	Statistic	df	Sig.
BT Summation score: BT1 to BT3	21-30	.240	89	<.001	.913	89	<.001
	31-40	.172	62	<.001	.927	62	.001
	41-50	.196	49	<.001	.896	49	<.001

a. Lilliefors Significance Correction

Screen 6.15 Test of normality

		Levene Statistic	df1	df2	Sig.
BT Summation score: BT1 to BT3	Based on Mean	6.192	2	197	.002
	Based on Median	5.987	2	197	.003
	Based on Median and with adjusted df	5.987	2	186.935	.003
	Based on trimmed mean	6.261	2	197	.002

Screen 6.16 Tests of homogeneity of variance

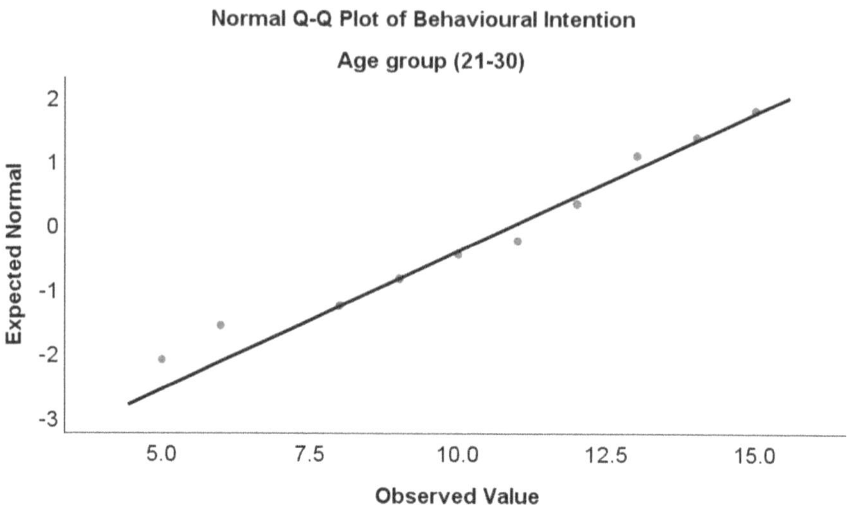

Screen 6.17 Normal Q-Q plot of behavioural intention 21 to 30 years old

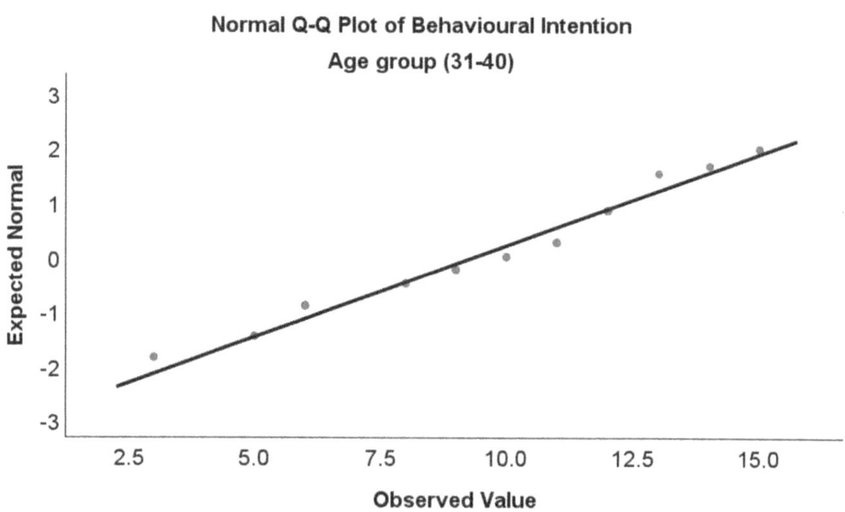

Screen 6.18 Normal Q-Q plot of behavioural intention 31 to 40 years old

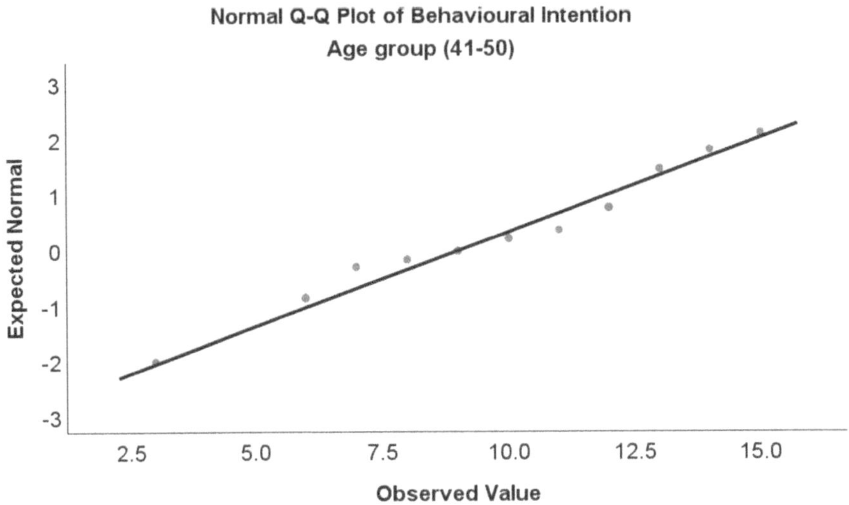

Screen 6.19 Normal Q-Q plot of behavioural intention 41 to 50 years old

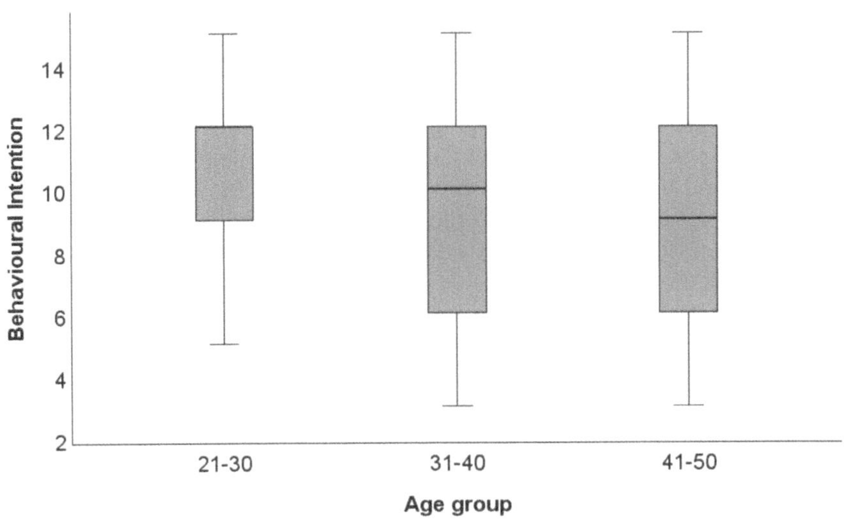

Screen 6.20 Boxplot of the age groups and behavioural intention

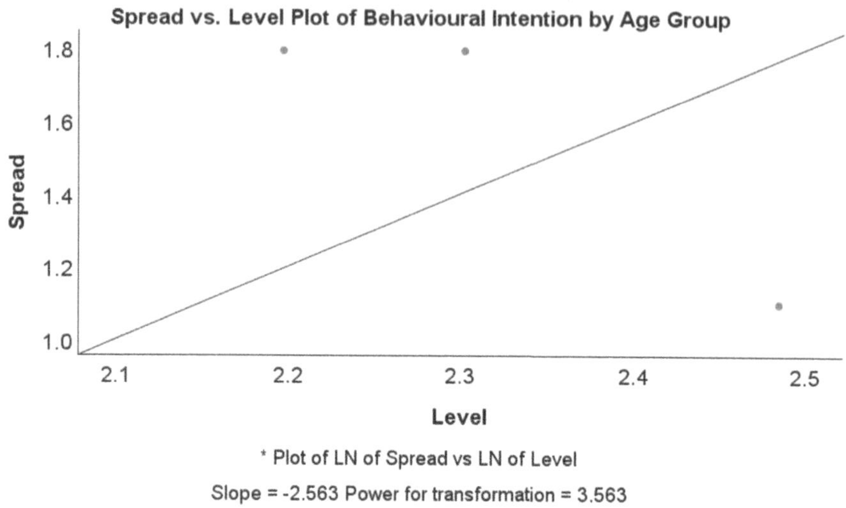

Screen 6.21 Spread vs. level plot

This analysis is **NOT** valid as both of the assumptions that are normality and equality of variance are not met.

The results of the test of normality ($KS = .000$, $p < 0.05$) and the Q-Q plot both indicate that the distribution of the mean behavioural intention for the age group 21 to 30 years old is asymmetrical (not normal). Likewise, the test of normality of the mean behavioural intention for age groups 31 to 40, and also 41 to 50, is both asymmetrical ($KS = .172$, $p < 0.05$; $KS = .196$, $p < 0.05$). The Q-Q plots do not show a normal distribution because some of the points do not fall along the diagonal line. On the other hand, the test of equality of variance indicates that the variances of the mean behavioural intention for different age groups are not equal. ($F = 6.192$, $p < 0.05$).

Recommendation

Since the data did not fulfil the normality assumptions, refer to the spread VS level plot, the power for estimation value obtained is 3.56, and it is suggested to use the cube transformation method.

Let's say the cube transformation is successful, then the parametric test (one-way ANOVA test) is appropriate. Otherwise, a nonparametric test such as Kruskal-Wallis would be more appropriate, which is equivalent to one-way ANOVA.

Table 6.1 Guidelines for data transformation if the data is not normal

Power	*Transformation*	*Description*
3	Cube	Each data value is cubed.
2	Square	Each data value is squared.
1	No change	No transformation or re-expression is required.
½	Square root	The square root of each data value is calculated.
0	Logarithm/natural log	Natural log transformation.
−½	Reciprocal of the square root	For each data value, the reciprocal of the square root is calculated.
−1	Reciprocal	The reciprocal of each data value is calculated.

6.4 Report writing of the one-way ANOVA test in APA format

A one-way ANOVA analysis of variance was performed to compare the behavioural intention of three age groups. The mean behavioural intention score for respondents by the identified age groups is shown in Figure 6.1. Before the analysis, assumption checking on both normality and homogeneity of variance was done for the mean behavioural intention score. The results of initial checking found that the two assumptions were not fulfilled and thus suggested cube transformation be performed. The results of one-way ANOVA revealed that there was a statistically significant difference in behavioural intention scores between three different age groups, F (2, 197) = 9.725, $p < 0.05$.

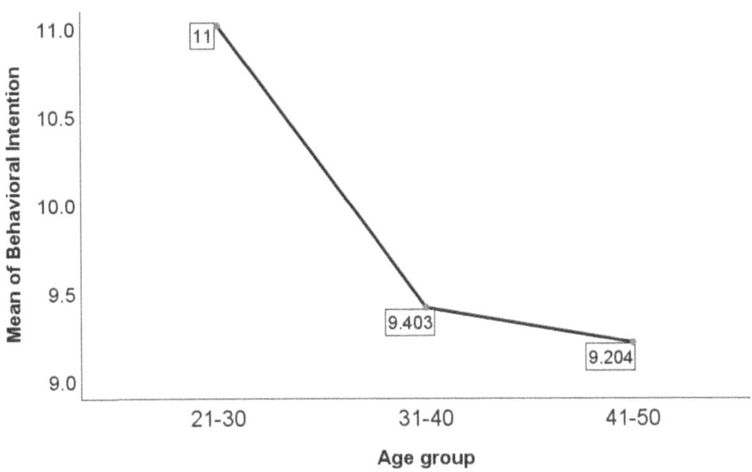

Figure 6.1 Mean score obtained by age groups and behavioural intention

Based on the Bonferroni Post Hoc multiple comparisons test, there was a statistically significant difference in the mean behavioural intention score for 21-to-30-year-olds (M = 11.00, SD = 2.32) and 31-to-40-year-olds (M = 9.40, SD = 2.97) with a large mean difference (MD = 1.60). Besides that, a statistically significant difference in the mean stress index score was also found in 21-to-30-year-olds (M = 11.00, SD = 2.32) and 41-to-50-year-olds (M = 9.20, SD = 2.95). On the other hand, no significant difference (MD = 0.20, p = .93) was noted for the mean behavioural intention score for 31-to-40-year-olds (M = 15.46, SD = 21.65) and 41-to-50-year-olds (M = 21.29, SD = 28.15). These findings indicate that the respondents in the age group of 21 to 30 years old were highly intended to use internet banking. Furthermore, the calculated eta-squared was 0.09. By referring to the effect size criteria set by Cohen (1988), 0.09 indicates that the mean difference of behavioural intention between age groups was large.

Present a summary table of ANOVA in APA format.

Table 6.2 One-way ANOVA between age groups and behavioural intention

Variables	n	M	SD	F	η^2
Age groups	89	11.00	2.32	9.725**	0.09
21–30	62	9.40	2.97		
31–40	49	9.20	2.95		
41–50					

*** Significant level at 0.01*

7 Chi-square test

Learning outcomes

At the end of this section, participants will be able to:

7.1 Recognize the purposes and requirements of chi-square analysis,
7.2 Know the three types of chi-square test,
7.3 Demonstrate the chi-square test of independence step by step using SPSS,
7.4 Interpret the selected outputs, and
7.5 Report chi-square test of independence in APA format.

7.1 Purpose and requirements of the chi-square test

Chi-square is used to identify the significant difference between the observed and expected data.

The requirements of both independent and dependent variables are categorical data; it can be a nominal or ordinal scale with two or more groups.

When a data set did not fulfil the normality assumption, the researcher opts to use a non-parametric test such as the chi-square test.

There are three (3) types of chi-square tests, the descriptions of which are as follows:

Goodness-of-fit (GOF)

It is used to check the difference between observed values and expected values that fit a certain predetermined table or matrix of the same size.

Homogeneity

This test aims to examine whether the frequency is equally distributed across the groups. In simple words, it compares the proportions of respondents' responses in categorical data. A contingency table (cross-tabulation) is used to display the information of the dependent variable in a row and the independent variable in a column.

DOI: 10.4324/9781003673903-7

Table 7.1 Example of research objectives that need a chi-square test

Research objective examples	*Independent variable*	*Dependent variable*
1. To identify the relationship between gender and online banking adoption	Gender (Male/Female)	Online banking adoption (Adopter/Non-adopter)
2. To identify the relationship between job category and job satisfaction level	Job category (Administrative, Sales, Managerial)	Job satisfaction level (Low, High)
3. To identify the association between income and happiness level	Income (<$1000, $1001–3000, $3000–5000, >$5000)	Happiness level (Low, Moderate, High)
4. To identify the relationship between the age group and behavioural intention level	Age group (21–30, 31–40, 41–50)	Behavioural intention level (Low, High)

Chi-square** (χ^2) **test of independence/association

This test is also known as a test that helps determine the statistical significance of the association or relationship between two variables with categorical data. Similarly, a contingency table of the selected variables will be presented.

Assumptions of the chi-square test

1. Random sampling: Each respondent has an equal chance of being selected from the population.
2. Independent observations: Every respondent can be counted one time only and cannot be counted again in other groups.
3. Expected frequency: At least 5 for the majority of the cells.

The next section contains instructions on the SPSS procedure to generate the chi-square test of independence. The following is the sample research objective for this test:

To identify the relationship between the age group and behavioural intention level.

7.2 Steps to run a chi-square test of independence using SPSS (example of 2×3 table)

1. On the menu bar, select **File**, **Open**, and **Data** (see Screen 7.1).

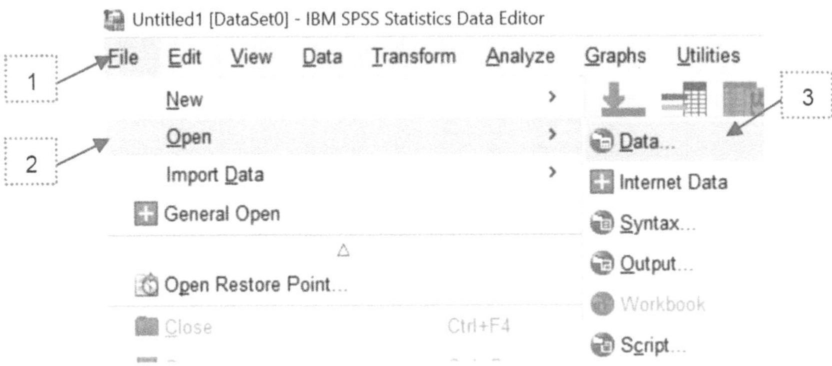

Screen 7.1 Data editor

2. In the sample data SPSS folder, select the **Internet Banking** data file. Click **Open** to open up the data file (see Screen 7.2).

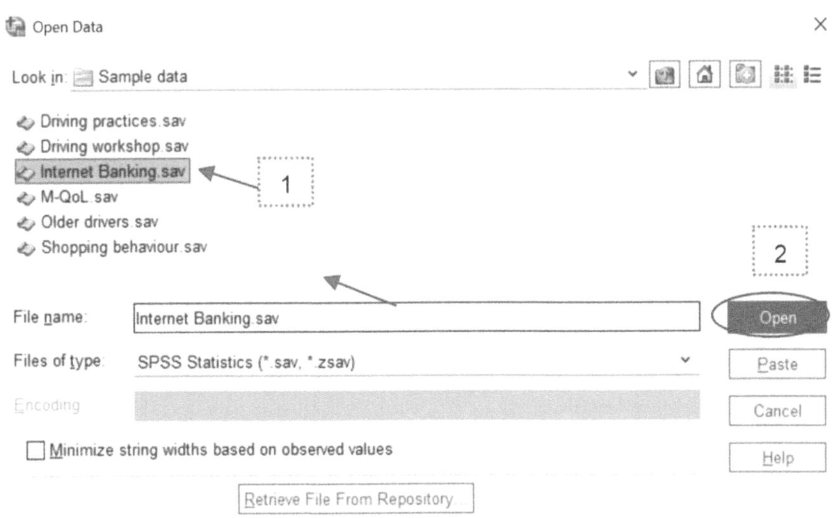

Screen 7.2 Open data

3. Go to the main menu, click **Analyze**, then go to **Descriptive Statistics** and select **Crosstabs** (see Screen 7.3).

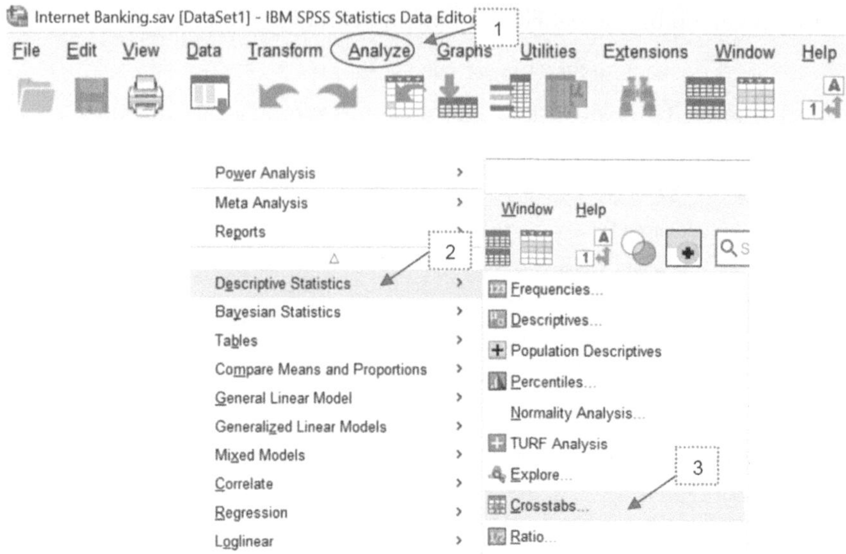

Screen 7.3 Select Crosstabs

4. From the **Variable** list box, select the dependent variable (**Internet Banking Adoption**) and move it to the **TargetList**. Next, select the independent variable (**Internet Banking Experience**) from the **Variabl**e list box and move it to the **Column(s)** box (see Screen 7.4).

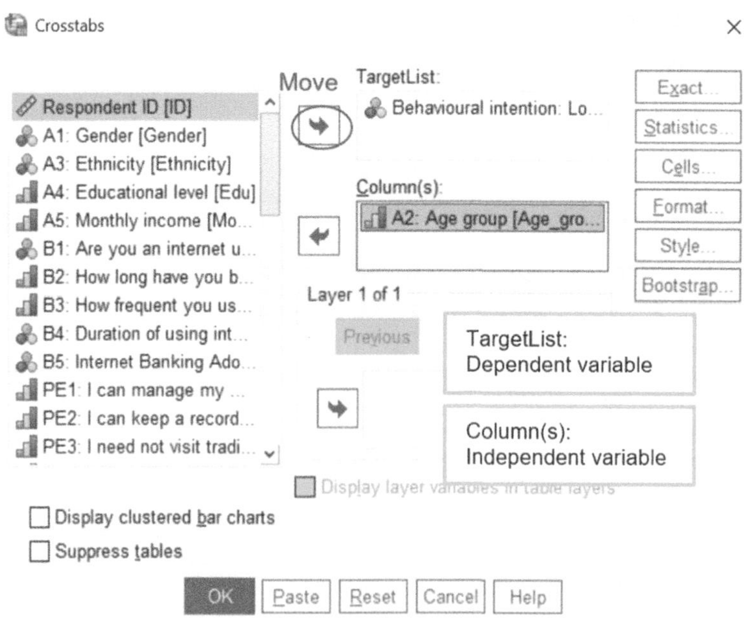

Screen 7.4 Crosstabs: Move variable

5. Click the **Statistics** button to open up the **Crosstabs: Statistics** options (see Screen 7.5).

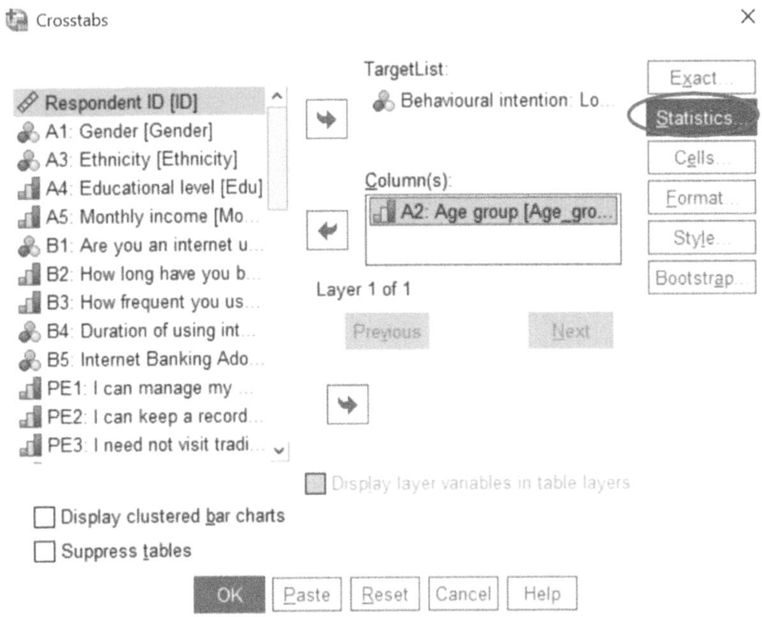

Screen 7.5 Crosstabs: Statistics

6. Check the options of **Chi-square**, **Contingency coefficient**, **Phi**, and **Cramer's V**. Then, click Continue and go back to **Crosstabs** (see Screen 7.6).

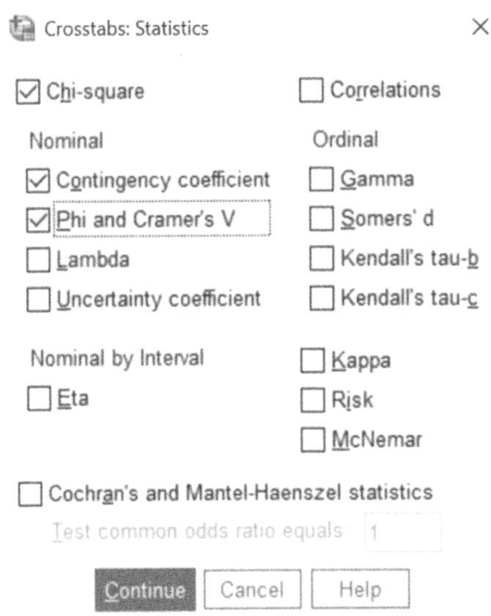

Screen 7.6 Select statistics options

7. Click **Cells** to open the **Crosstabs: Cells Display** option (see Screen 7.7).

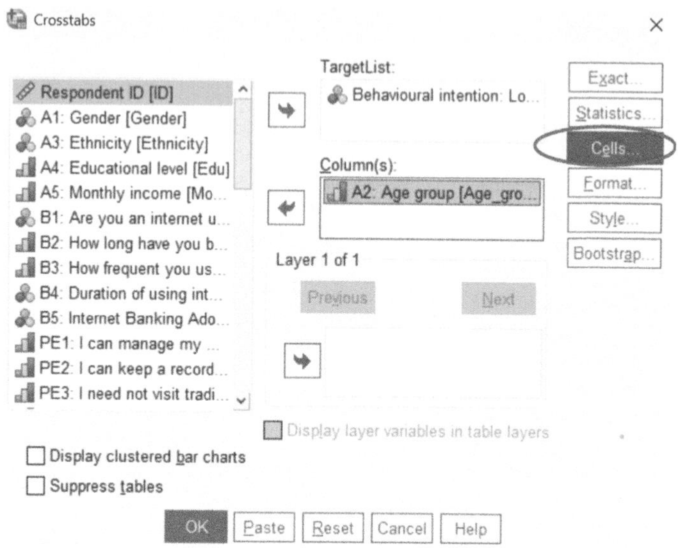

Screen 7.7 Crosstabs: Cells Display

8. See Screen 7.8. Check **Expected**, under **Counts**, and both Percentages of **Row** and **Total**, followed by **Continue**. Remember to select the option to **Create APA-style table**. Lastly, click **OK** to run the test in Screen 7.9.

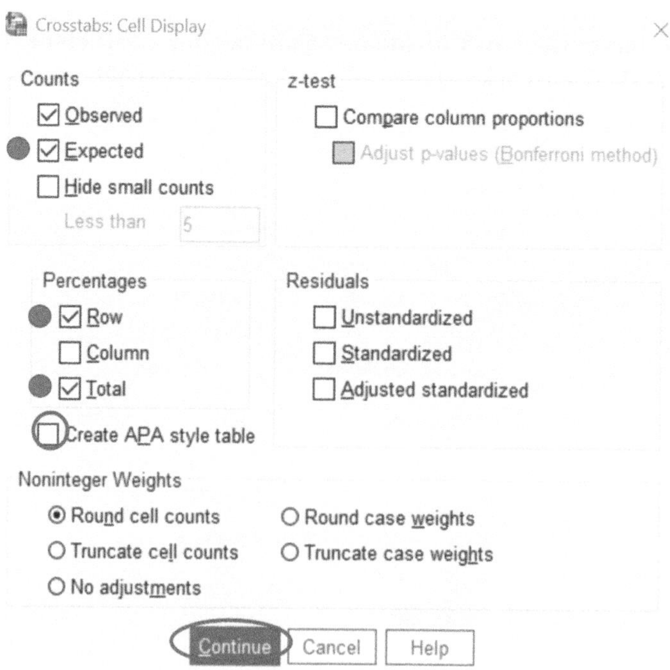

Screen 7.8 Cells Display options

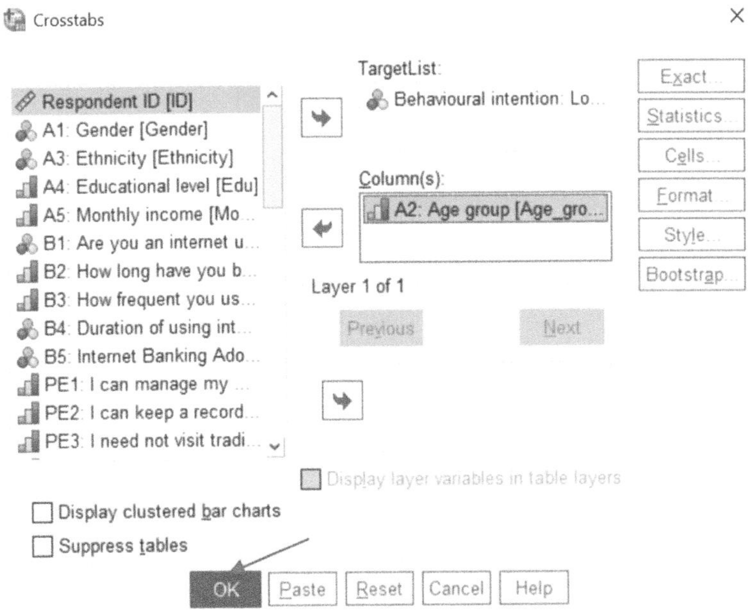

Screen 7.9 Run crosstabs

7.3 Hypothesis testing and interpretation of the chi-square test of independence

Given that, the researcher would like to examine whether there is any relationship between **age group** and **behavioural intention level**. Based on the purpose of the research objective and data scale measurement, a chi-square test of independence would be appropriate. Screens 7.10–7.12 are the generated SPSS outputs.

Screen 7.11 displays the SPSS output if the option of APA format style table is selected.

Behavioural intention: Low/ High * A2: Age group Crosstabulation

			A2: Age group			
			21-30	31-40	41-50	Total
Behavioural intention: Low/ High	Low (Lowest to 11)	Count	38	42	31	111
		% within Behavioural intention: Low/ High	34.2%	37.8%	27.9%	100.0%
		% of Total	19.0%	21.0%	15.5%	55.5%
	High (12 to Highest)	Count	51	20	18	89
		% within Behavioural intention: Low/ High	57.3%	22.5%	20.2%	100.0%
		% of Total	25.5%	10.0%	9.0%	44.5%
Total		Count	89	62	49	200
		% within Behavioural intention: Low/ High	44.5%	31.0%	24.5%	100.0%
		% of Total	44.5%	31.0%	24.5%	100.0%

Screen 7.10 Cross-tabulation

Behavioural intention: Low/ High * A2: Age group Crosstabulation

		A2: Age group							
		21-30		31-40		41-50		Total	
		N	%	N	%	N	%	N	%
Behavioural intention: Low/ High	Low (Lowest to 11)	38	42.7%	42	67.7%	31	63.3%	111	55.5%
	High (12 to Highest)	51	57.3%	20	32.3%	18	36.7%	89	44.5%
Total		89	100.0%	62	100.0%	49	100.0%	200	100.0%

Screen 7.11 Cross-tabulation in APA format

	Value	df	Asymptotic Significance (2-sided)
Pearson Chi-Square	10.866[a]	2	.004
Likelihood Ratio	10.951	2	.004
Linear-by-Linear Association	7.160	1	.007
N of Valid Cases	200		

a. 0 cells (0.0%) have expected count less than 5. The minimum expected count is 21.81.

Symmetric Measures

		Value	Approximate Significance
Nominal by Nominal	Phi	.233	.004
	Cramer's V	.233	.004
	Contingency Coefficient	.227	.004
N of Valid Cases		200	

Screen 7.12 Chi-square tests

A chi-square test is considered **valid** if the percentage of expected count (N) is less than 5 is ≤20%.

In this case, all expected count is more than 20.

1. State the alternative hypotheses based on the research objective.
   The alternate hypothesis (H_a):
   There is a relationship between age group and behavioural intention level.
   or
   Age group is dependent on the behavioural intention level.
2. Based on the SPSS output, decide on the null hypothesis by using a 0.05 level of significance and conclude.
   The obtained p-value (p = .004) is very much smaller than 0.05. Therefore, H_a is accepted. Therefore, we can conclude that there is a significant relationship between age group and behavioural intention level.
   or
   Age group is dependent on the behavioural intention level.

Table 7.2 Chi-square test of independence for age group and behavioural intention level

Behavioural intention level	*Age group*			*Total*
	21–30	*31–40*	*41–50*	
Low	38[a]	42	31	111
	34.2[b]	37.8	27.9	55.5
High	51	20	18	89
	57.3	22.5	20.2	44.5
Total	89	62	49	200
	44.5	31.0	24.5	100.0

Pearson chi-square = 10.87 *df* = 2 *P*-value = 0.004
Cramer's *V* = 0.23, % cell with expected count less than 5 = 0.0%
[a] = Count/frequency
[b] = Row %/Within level %

3. Describe the trend and magnitude of the relationship, if there is any.

   The distribution of the level of perceived behavioural intention to adopt internet banking was almost equally distributed. Generally, respondents aged between 21 and 30 years had perceived higher behavioural intention towards (57.3%) as compared to those aged 31 to 40 years (22.5%) and aged 41 to 50 years (20.2%).

   This reflects that those who are aged 21 to 30 years had a significantly higher proportion of adopting internet banking services.

   The respondents with a low level of behavioural intention were distributed about 30% in each group, and this shows no obvious difference.

   According to Cohen (1988), the magnitude of the relationship in Cramer's V is as follows:

   Small effect: Cramer's V ≈ 0.1
   Medium effect: Cramer's V ≈ 0.3
   Large effect: Cramer's V ≈ 0.5

   Refer to SPSS outputs, under the symmetric measures in Screen 7.12, the value of <u>Cramer's V</u> was <u>0.23. This indicated that</u> the <u>magnitude or strength</u> of the relationship was a small to medium effect.
4. Justify whether this chi-square analysis is valid or appropriate. If it is not valid, provide a recommendation.

This analysis is valid because the percentage of the cells with the expected count that was less than 5 was 0.0%. Hence, it is not necessary to combine cells.

7.4 Sample write-up of the chi-square test of independence

A chi-square test of independence was conducted to examine whether the age group is related to the behavioural intention to use internet banking. The results show that age group is related to perceived behavioural intention to adopt internet banking [χ^2 (2, N = 200) = 10.87, $p < 0.05$].

Table 7.3 Chi-square test of independence for age group (IV) by perceived behavioural intention to adopt internet banking (DV)

Behavioural intention level	*Age group*			*Total*
	21–30	*31–40*	*41–50*	
Low	38[a]	42	31	111
	34.2[b]	37.8	27.9	55.5
High	51	20	18	89
	57.3	22.5	20.2	44.5
Total	89	62	49	200
	44.5	31.0	24.5	100.0

Pearson chi-square = 10.87 $df = 2$ P-value = 0.004
Cramer's $V = 0.23$, % cell with expected count less than 5 = 0.0%
[a] = Count/frequency
[b] = Row %/Within level %

A close examination of Table 7.3 to check the pattern of the relationship was done. The distribution of the level of perceived behavioural intention to adopt internet banking was almost equally distributed. Generally, respondents aged between 21 and 30 years had perceived higher behavioural intention (57.3%) as compared to those aged 31 to 40 years (22.5%) and aged 41 to 50 years (20.2%).

This reflects that those who are aged 21 to 30 years had a significantly higher likelihood of adopting internet banking services. The respondents with a low level of behavioural intention were almost distributed about 30% in each group, and this shows no obvious difference.

Based on the Cramer's V value ($V = 0.23$), the relationship between age group and behavioural intention to use internet banking was considered small to medium. The percentage of cells with an expected count of less than 5 did not exist. This shows that the conducted analysis was appropriate, as the chi-square test assumption was met.

8 Correlation

Learning outcomes

At the end of this section, participants will be able to:

8.1 Recognise the purposes and requirements of correlation analysis,
8.2 Demonstrate correlation step by step using SPSS,
8.3 Interpret the correlation outputs, and
8.4 Report correlation in APA format.

8.1 Purpose and requirements of correlation analysis

Correlation analysis is also known as Pearson's correlation or Pearson product-moment correlation. It is used to determine the relationship between *two continuous or metric variables.* In other words, the requirement of both independent and dependent variables is the scale of measurement of interval or ratio (continuous data). To run a correlation analysis, the variables need to fulfil the assumption of normality, linearity, and homoscedasticity.

Two important components of correlation, r

1. Direction

   The $\boldsymbol{r}$ value ranges from $-1 \leq \boldsymbol{r} \leq 1$.

   It is possible to have a positive ($+r$ value), negative ($-r$ value), or no ($r = 0$) relationship among the selected variables. Positive correlation ($0 < r < 1$) means that when the independent variable increases, the dependent variable will also increase. For example, a study relevant to weight management wants to know the relationship between the daily workout time in hours and the monthly weight loss in kg. The data of the participants has been collected and reflects a positive correlation as depicted in Figure 8.1.

   If the relation of both independent variables has a constant change in a positive trend, this is called perfect positive correlation ($r = 0$). Let's say, a household's electric consumption and the monthly electric bill. The electric expense rate has a fixed rate, and the sample of pattern data distribution is displayed in Figure 8.2.

DOI: 10.4324/9781003673903-8

Figure 8.1 Positive correlation

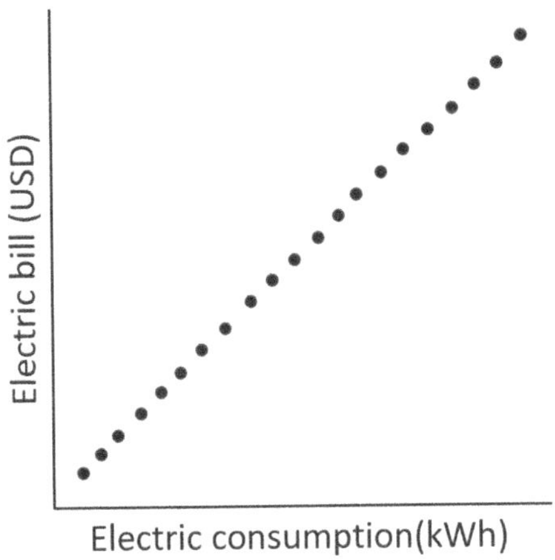

Figure 8.2 Perfect positive correlation

In contrast, a negative correlation ($-1 < r < 0$) means that when the independent variable increases, the dependent variable will decrease (see Figure 8.3). For example, the body weight of the participants was measured in kilograms, and their self-confidence score was also measured. This means that if the body weight of the individual is higher than average, she may rate her self-confidence lower. Perfect negative correlation ($r = -1$) is another scenario when both independent variables have a constant change in a negative trend. The change at a fixed rate may be represented by the balance of petrol in litres and distance travelled in kilometres of a car, as shown in Figure 8.4.

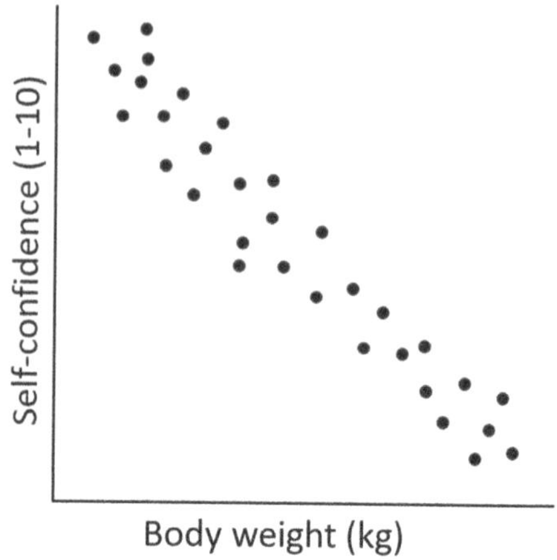

Figure 8.3 Negative correlation

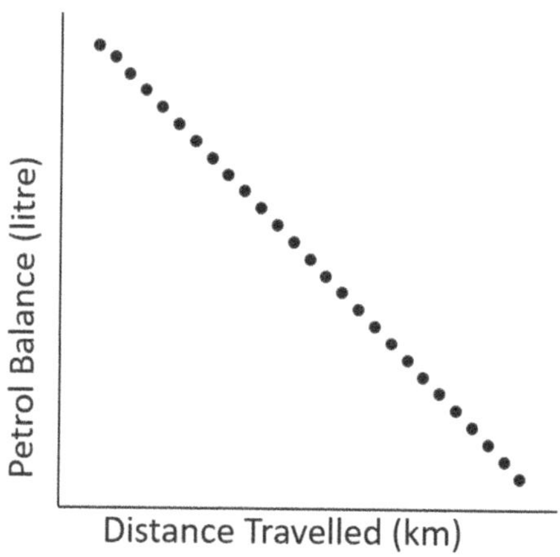

Figure 8.4 Perfect negative correlation

In the case of no correlation (*r* = 0), the variables do not bring change to each other, such as the number of cups of coffee consumed daily by the students and their CGPS score (see Figure 8.5).

2. Strength/Magnitude

   The ***r*** value close to 0 means a weak relationship, and close to 1 means a stronger relationship. Refer to the Guilford rule of thumb (Guilford, 1950); the magnitude of the *r* value means the relationship is:

   <.2 Negligible
   .2–.4 Low
   .4–.7 Moderate
   .7–.9 High
   >.9 Very high

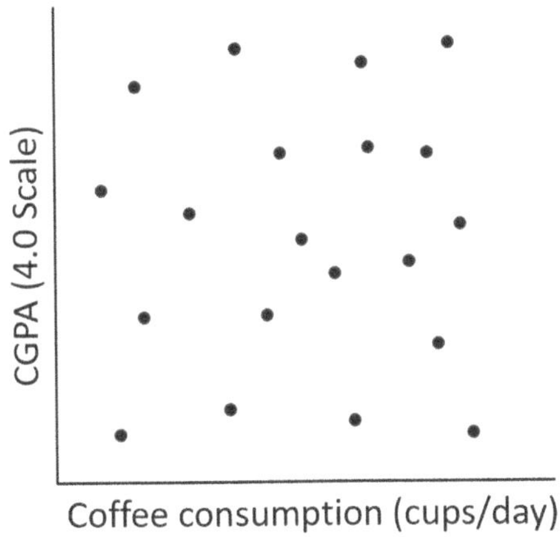

Figure 8.5 No correlation

Tips for a successful correlation analysis

Transformation is needed for a multi-item scale (e.g. summation score, index). Remember to compute the score, and also be mindful of the negative statements. Test the assumptions of correlations before performing the correlation. For more details on how to do data transformation, refer to Chapter 2.

Working example

The working data of internet banking is part of the study from Yeoh and Chan (2011).

Sample research objective:

To identify the relationship between performance expectancy (PE) and behavioural intention (BI) to use internet banking among Kuala Lumpur residents.

Independent variable: performance expectancy
Dependent variable: behavioural intention

* Both variables were collected using multiple questions with 5-point Likert scales. Hence, before the correlation analysis, a transformation needs to be done to compute the summation score. Sections 8.1.1 and 8.1.2 focus on assumption testing.

8.1.1 Testing the assumption of normality

To run an EDA analysis, refer to Chapter 1. In step 4, insert the performance expectancy (PE) summation score and behavioural intention (BI) summation score into the dependent variable.

Firstly, check the formal normality statistical tests (Kolmogorov-Smirnov or Shapiro-Wilk test). Alternatively, check for data skewness (close to 0, data is normal) for data distribution and kurtosis of the height and sharpness of the central peak of the data. Generally, the acceptable range of skewness and kurtosis is between −2 and +2 (George & Mallery, 2024). The obtained skewness for PE is −0.372, and kurtosis is −0.082; skewness for BI is −0.488, and kurtosis is −0.551. Therefore, the data have met normal distribution.

Secondly, evaluate the normality of the data through graphical methods (i.e. histogram, normality plot). As depicted in the histogram (see Figures 8.6–8.7),

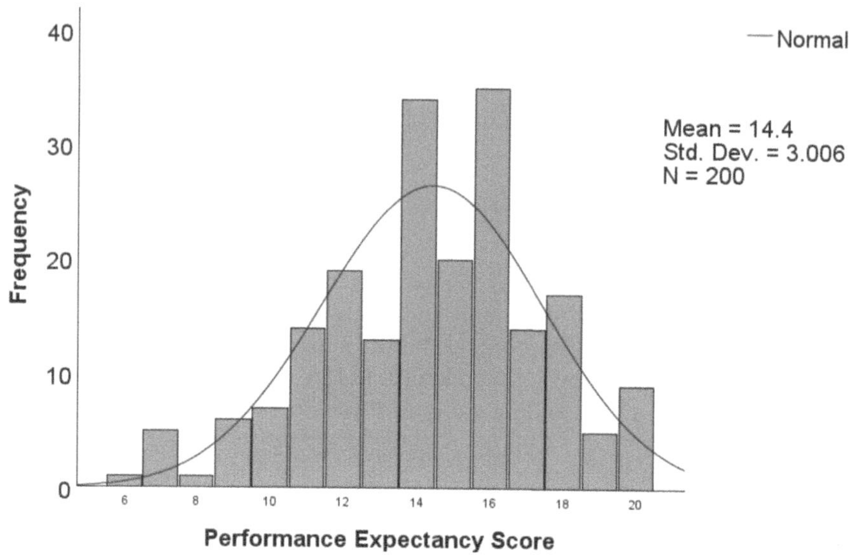

Figure 8.6 Histogram of performance expectancy score

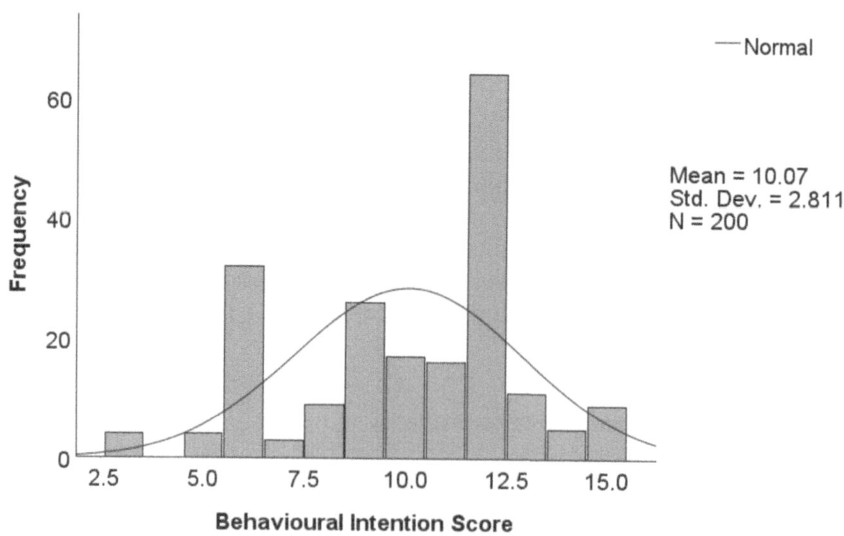

Figure 8.7 Histogram of behavioural intention score

the data distribution seems to fall under the bell curve. Whereas, the data falling along or near the normal Q-Q plot is evidence that the data fulfilled the normality assumption (see Figures 8.8–8.9).

Refer to the boxplot in Figure 8.10 and Figure 8.11; there are no outliers or extreme scores. If an outlier exists, a small circle and an asterisk (*) with the respective data file row number will be shown.

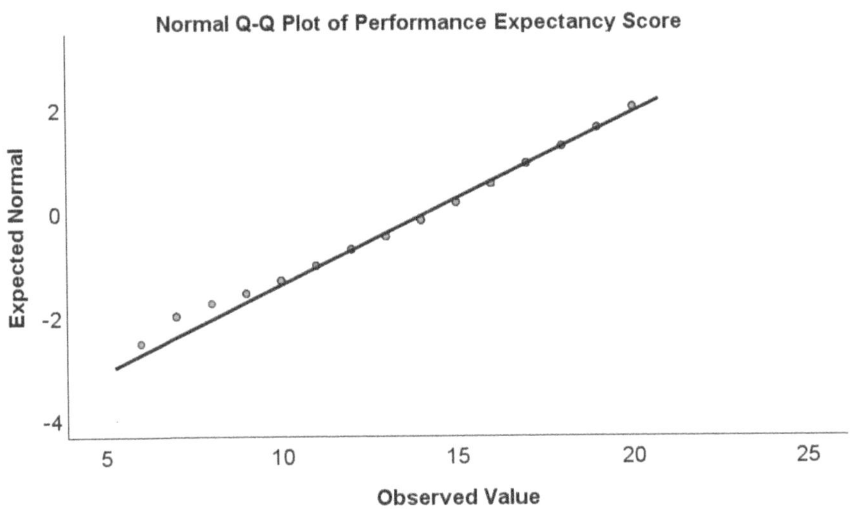

Figure 8.8 Normal Q-Q plot of performance expectancy score

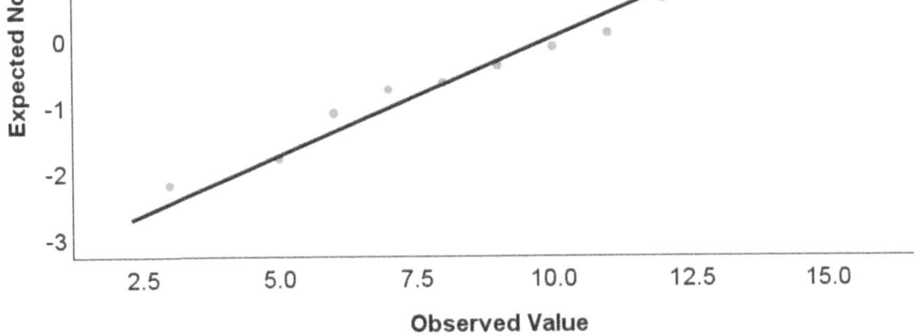

Figure 8.9 Normal Q-Q of behavioural intention score

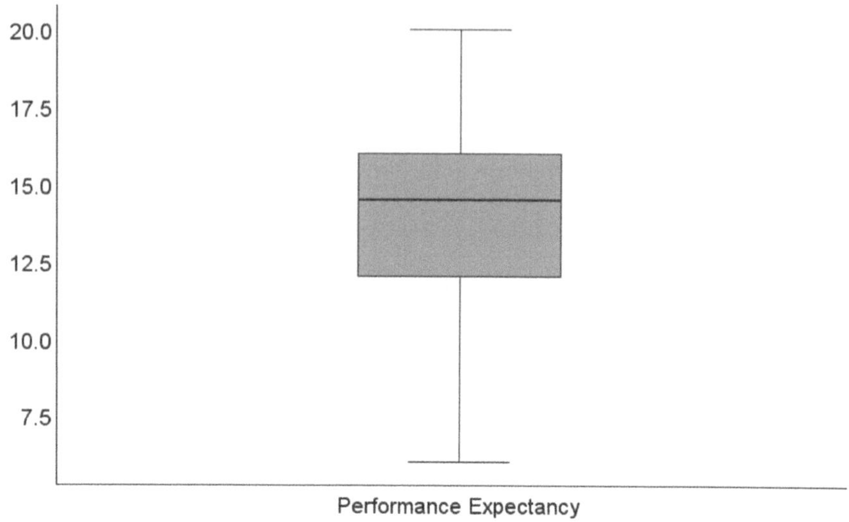

Figure 8.10 Boxplot of performance expectancy score

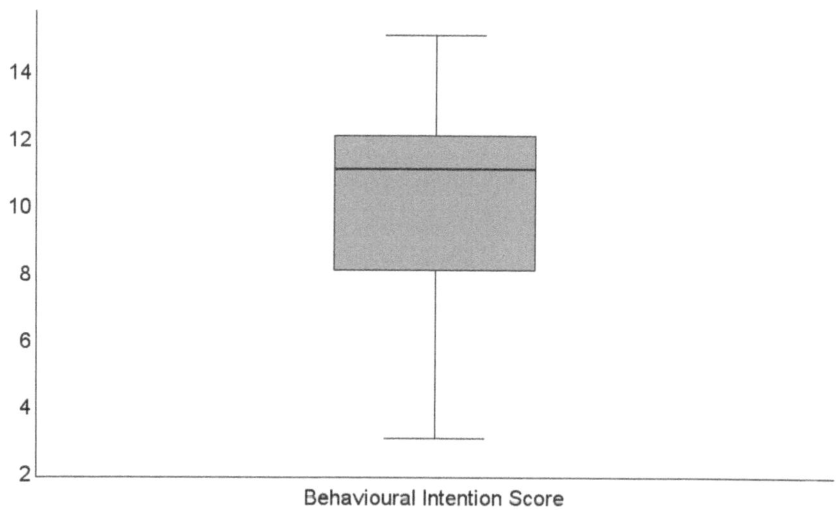

Figure 8.11 Boxplot of behavioural intention score

8.1.2 Testing the assumption of linearity

A linearity test needs to be conducted to test the assumption. The relationship between the independent variables is linearly dependent when the value sig. Deviation from Linearity is larger than 0.05 (>0.05), or vice versa.

Steps to perform linearity tests

1. On the menu bar, select **File**, **Open**, and **Data** (see Screen 8.1). The Open dialogue box appears.

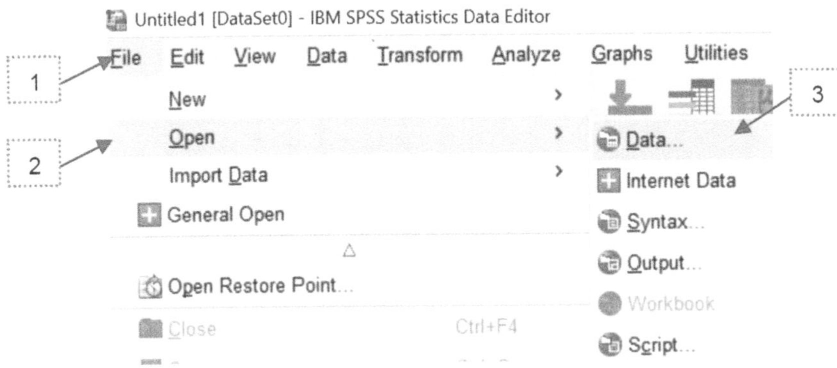

Screen 8.1 Data editor

2. In the sample data SPSS folder, select the **Internet Banking** data file and click **Open** to open the data file (see Screen 8.2).

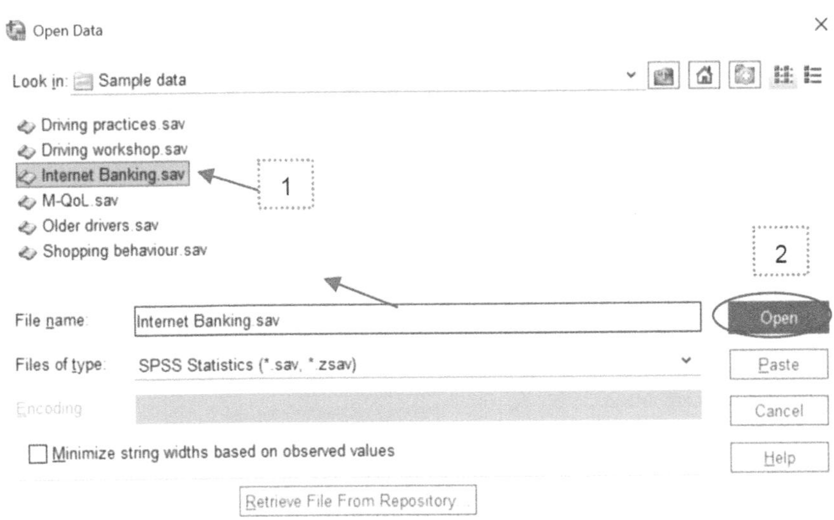

Screen 8.2 Open data

3. Go to the main menu, click **Analyze** followed by **Compare Means,** and select **Means** (see Screen 8.3).

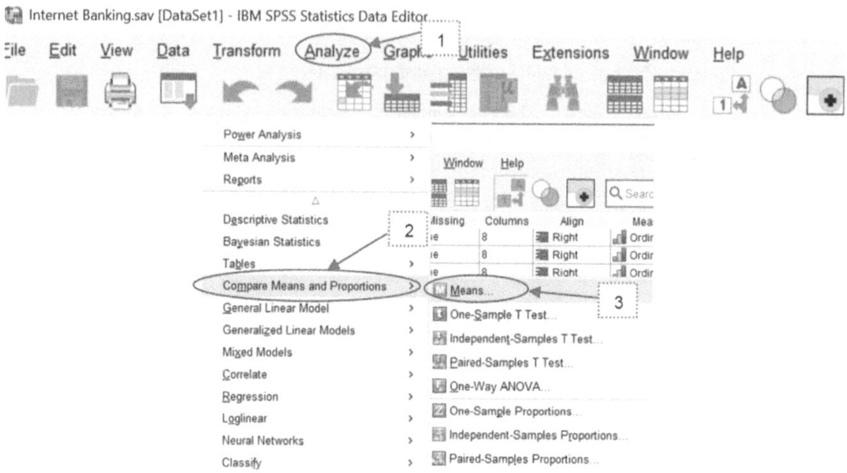

Screen 8.3 Select Means

4. From the variable list box, select the independent variable (**Behavioural Intention summation score**) and enter it into the **Dependent List** box. Next, select the independent variable (**Performance Expectancy summation score**) and move it to the **Independent List** box (Screen 8.4).

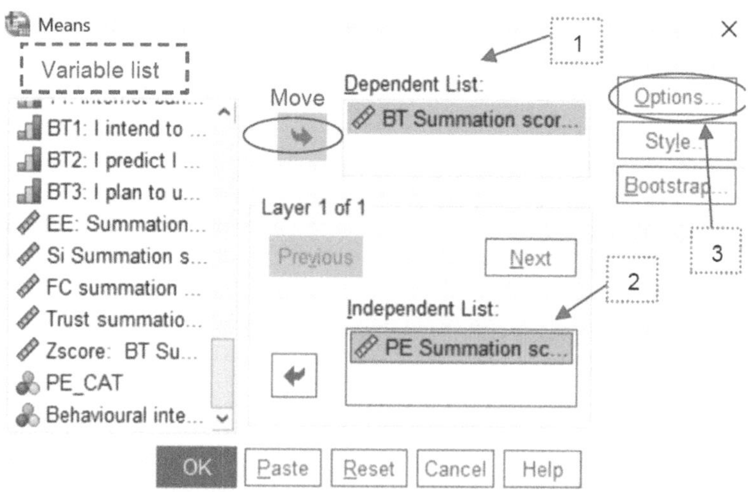

Screen 8.4 Means: Insert variable

5. Continuing Screen 8.4, click **Options**. Under **Statistics for First Layer**, check ☑ **Test for Linearity**, and then click **Continue** (see Screen 8.5).

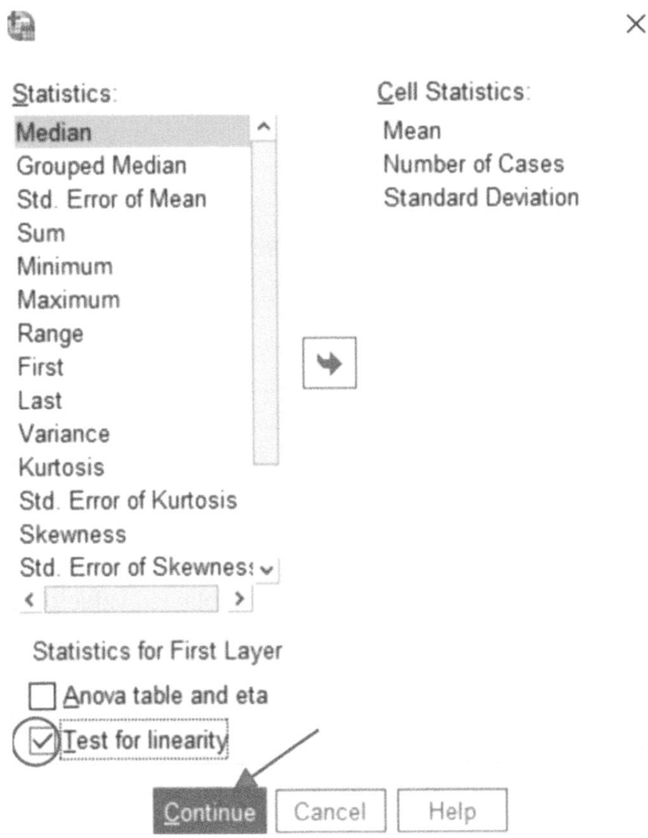

Screen 8.5 Means: Options

6. Click **OK** in Screen 8.6 to run the test.

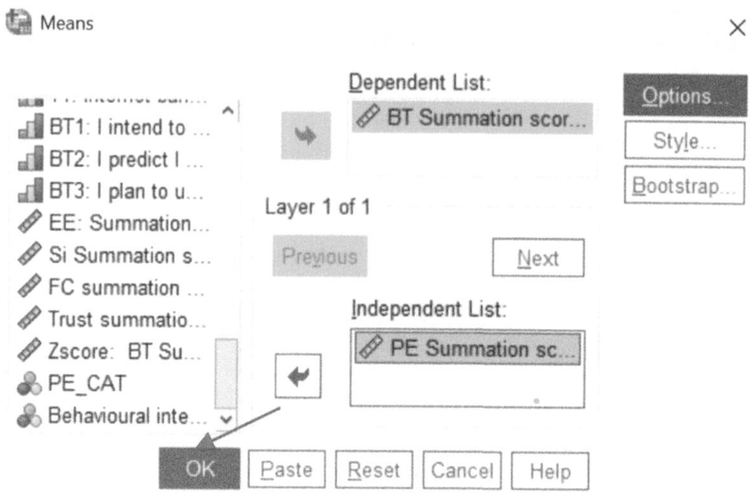

Screen 8.6 Run the test

Selected output of the linearity test and its interpretation

Based on the ANOVA table in Screen 8.7, the obtained value of sig. Deviation from Linearity is 0.240 (>0.05). Thus, the conclusion is that there is a linear relationship between the variables of PE and BI. Therefore, the linearity assumption is fulfilled.

Since both assumptions of correlation are fulfilled. Thus, we can proceed to correlation analysis now.

			Sum of Squares	df	Mean Square	F	Sig.
BT Summation score* PE Summation score	Between Groups	(Combined)	503.999	14	36.000	6.235	<.001
		Linearity	409.353	1	409.353	70.898	<.001
		Deviation from Linearity	94.647	13	7.281	1.261	.240
	Within Groups		1068.156	185	5.774		
	Total		1572.155	199			

Measures of Association

	R	R Squared	Eta	Eta Squared
BT Summation score * PE Summation score	.510	.260	.566	.321

Screen 8.7 ANOVA table

8.2 Steps to run correlation analysis

1. On the menu bar, select **File**, **Open**, and **Data** (see Screen 8.8). The open data window will appear.

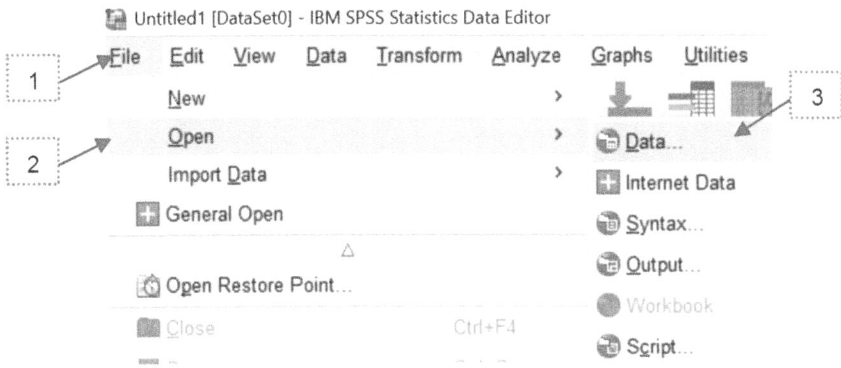

Screen 8.8 Data editor

2. In the sample data SPSS folder, select the **Internet Banking** data file and click **Open** to open the data file (see Screen 8.9).

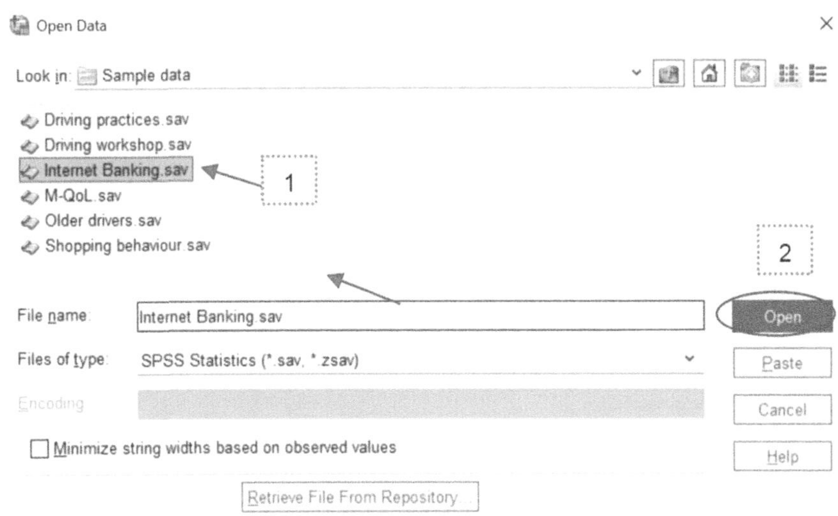

Screen 8.9 Open data

3. Go to the main menu, choose **Analyze**, followed by **Correlate**, and click **Bivariate** (see Screen 8.10).

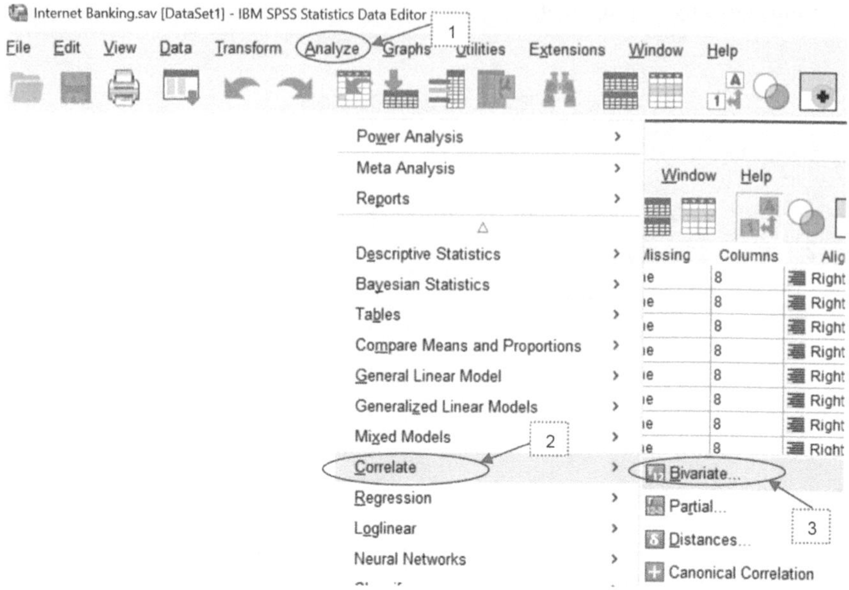

Screen 8.10 Choose Analysis

4. From the variable list box in Screen 8.11, select the independent variable (**Performance Expectancy summation score**) and dependent variable (**Behavioural Intention summation score**). Then, enter them into the **Variables** box (see Screen 8.12).

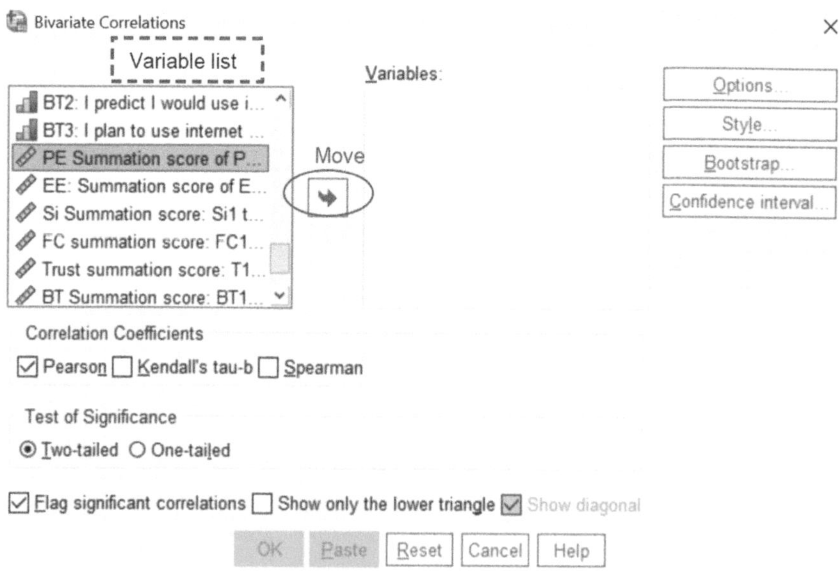

Screen 8.11 Bivariate Correlations

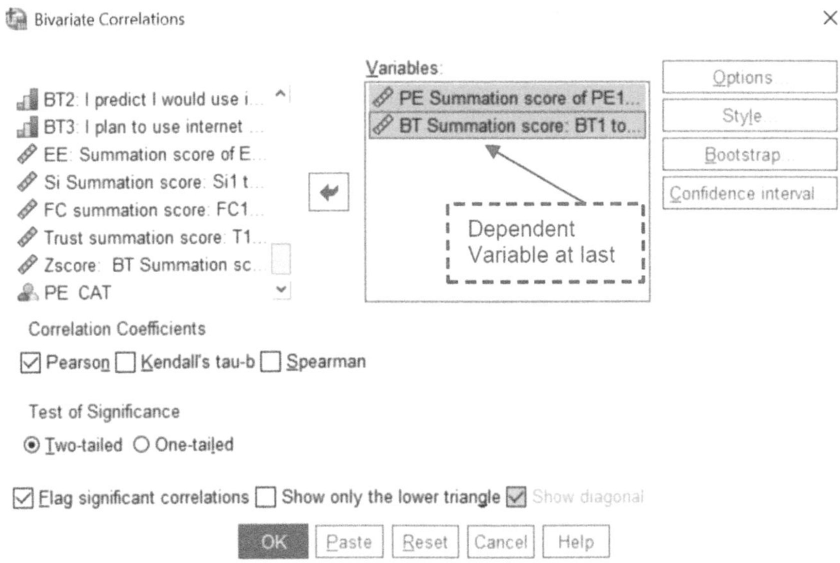

Screen 8.12 Insert variables

5. Click **Options**, open the dialogue box Statistics, and select the **Means and standard deviations** in Screen 8.13.

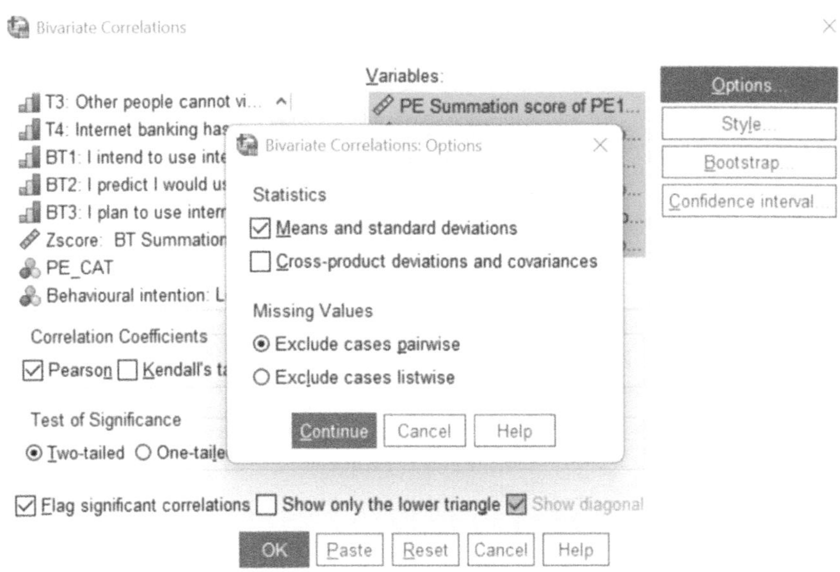

Screen 8.13 Bivariate Correlations: Options

6. Click the **Confidence interval** option to select **Estimate confidence interval of bivariate correlation parameter** in Screen 8.14. System default runs a confidence interval at 95%.

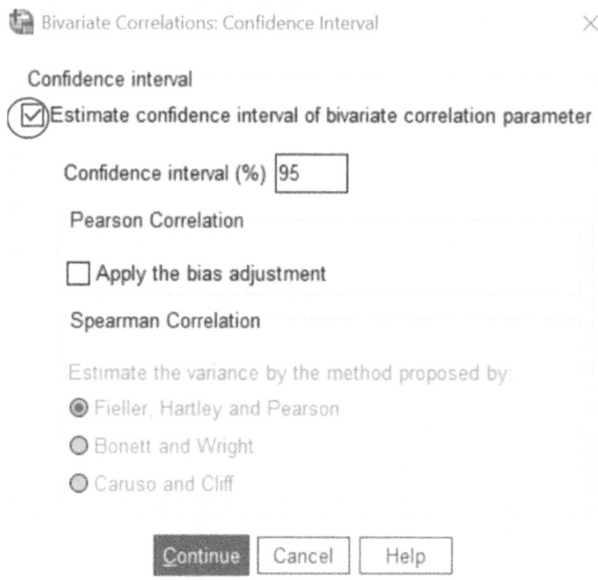

Screen 8.14 Bivariate Correlations: Confidence Interval

7. Lastly, click **OK** in Screen 8.15 to run the analysis.

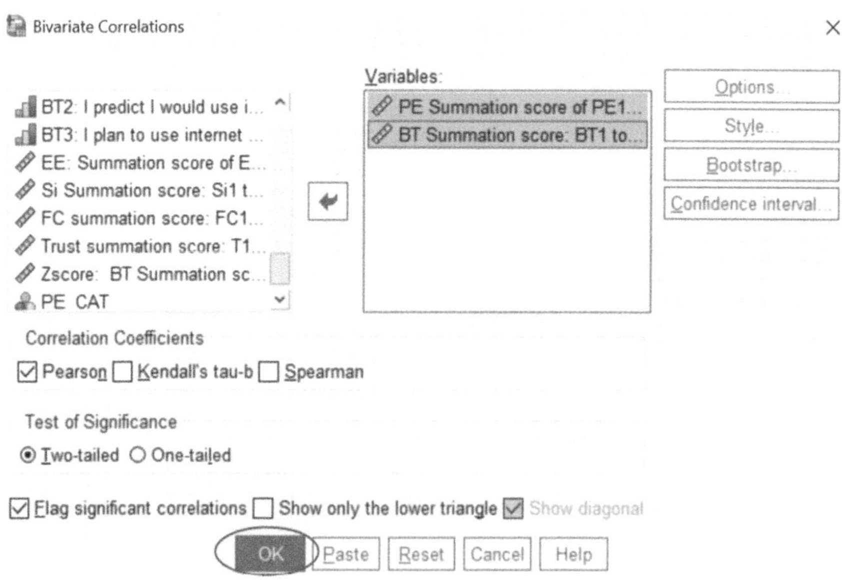

Screen 8.15 Run the analysis

8.3 SPSS output of correlation analysis

Descriptive Statistics

	Mean	Std. Deviation	N
Performance Expectancy Summation score	14.40	3.006	200
Behavioural Intention Summation score	10.07	2.811	200

Correlations

		Performance Expectancy Summation score	Behavioural Intention Summation score
Performance Expectancy Summation score	Pearson Correlation	1	.510
	Sig. (2-tailed)		<.001
	N	200	200
Behavioural Intention Summation score	Pearson Correlation	.510**	1
	Sig. (2-tailed)	<.001	
	N	200	200

**. Correlation is significant at the 0.01 level (2-tailed).

r valu

Confidence Intervals

	Pearson Correlation	Sig. (2-tailed)	95% Confidence Intervals (2-tailed)[a] Lower	Upper
Performance Expectancy Summation score Behavioural Intention Summation score	.510	<.001	.400	.606

a. Estimation is based on Fisher's r-to-z transformation.

Screen 8.16 Descriptive Statistics, Correlations and Confidence Intervals

8.4 Hypothesis testing and interpretation of correlation SPSS output

Refer to the research objective in Chapter 8.1, and the SPSS outputs in Screen 8.16.

1. State the null and alternative hypotheses based on the research objective.
   Null hypothesis
   H_0: There is no relationship between performance expectancy and behavioural intention to use internet banking among Kuala Lumpur residents.
   Alternative hypothesis
   H_a: There is a relationship between performance expectancy and behavioural intention to use internet banking among Kuala Lumpur residents.
2. Based on Screen 8.16, decide on the null hypothesis by using a 0.05 level of significance and conclude.
   Decision: Null hypothesis (H_0) is rejected, and therefore, H_a is supported.

Note that the corresponding p-value is smaller (p = .0001) than the .05 level of significance. Therefore, there is a relationship between performance expectancy and behavioural intention to use internet banking.

3. Describe the trend and magnitude of the relationship, if there is any.

   The obtained correlation coefficient was 0.510, which indicates the direction of the relationship is positive, with a moderate strength of the relationship between the selected variables.

4. Based on a 95% confidence interval, what is the true population correlation coefficient range for the performance expectancy and behavioural intention to use internet banking?

A 95% CI of [0.400, 0.606] of this correlation would indicate a significant positive correlation, because the value 0 is not within this interval. This finding indicates a 95% confidence that the true population correlation coefficient ranges between 0.400 and 0.606 for performance expectancy and behavioural intention.

8.5 Report correlation analysis in APA format

The relationship between performance expectancy and behavioural intention to use internet banking among Kuala Lumpur residents was examined using correlation coefficients. Preliminary analyses were conducted to ensure no violation of the normality and linearity assumptions. The correlation analysis results are shown in Table 8.1. This study found that there was a significant relationship between performance expectancy and behavioural intention (r = 0.510, p < 0.05). The strength of the relationship between variables is moderate. The positive correlation coefficient (r = 0.510) means that when performance expectancy increases, the behavioural intention also increases. Furthermore, a 95% confidence interval was computed of [0.400, 0.606]. This finding indicates a 95% confidence that the true population correlation coefficient ranges between 0.400 and 0.606 for performance expectancy and behavioural intention.

Table 8.1 Mean, standard deviation, Pearson's correlation coefficients, and confidence interval of the selected variables with behavioural intention to use internet banking

Variables	M	SD	r
Behavioural intention	14.40	3.01	
Performance expectancy	10.07	2.82	0.510** [0.400, 0.606]

M = mean, *SD* = standard deviation, *r* = Pearson's correlation coefficient, 95% confidence interval values were in brackets. **Correlation is significant at the 0.01 level (two-tailed).

> *Question*
>
> What about multiple independent variables with a dependent variable?
> Refer to Chapter 8.2, then repeat the correlation analysis steps 1 to 3. Then, continue step 4 to include the multiple independent variables.

4. Select the independent variable (**Performance Expectancy summation score . . .**) and the dependent variable (**Behavioural Intention summation score**) from the variable list box, enter them into the **Variable** box (Screen 8.17). Click **OK** to run the multiple correlations.

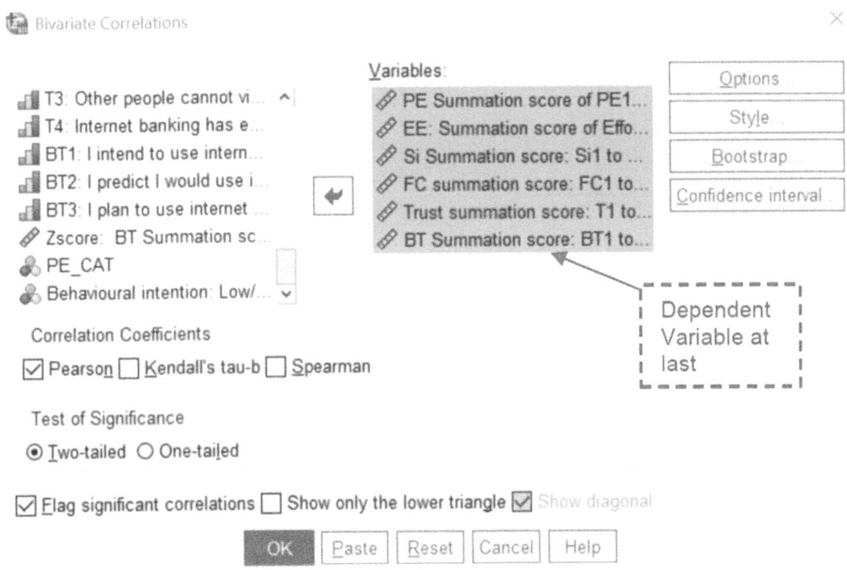

Screen 8.17 Bivariate Correlations

SPSS output of the correlations in Screen 8.18.

From Screen 8.18, the results of the correlation matrix are diagonally the same. Refer to the Pearson correlation value and the significant two-tailed p-value in the last column from the right for the correlation with the dependent variable. To prepare a correlation matrix, refer to the statistics either below or above the diagonal line.

Generally, before a multiple linear regression analysis takes place, a correlation analysis involving all of the selected variables will be performed. It is important to observe the relationship from the correlation matrix for identifying the potential multicollinearity.

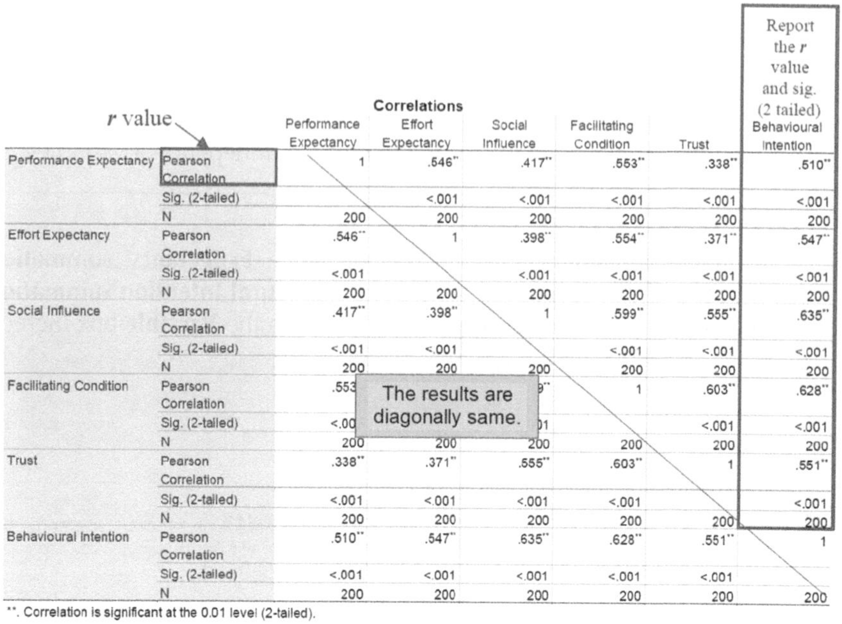

Correlations

		Performance Expectancy	Effort Expectancy	Social Influence	Facilitating Condition	Trust	Behavioural Intention
Performance Expectancy	Pearson Correlation	1	.546**	.417**	.553**	.338**	.510**
	Sig. (2-tailed)		<.001	<.001	<.001	<.001	<.001
	N	200	200	200	200	200	200
Effort Expectancy	Pearson Correlation	.546**	1	.398**	.554**	.371**	.547**
	Sig. (2-tailed)	<.001		<.001	<.001	<.001	<.001
	N	200	200	200	200	200	200
Social Influence	Pearson Correlation	.417**	.398**	1	.599**	.555**	.635**
	Sig. (2-tailed)	<.001	<.001		<.001	<.001	<.001
	N	200	200	200	200	200	200
Facilitating Condition	Pearson Correlation	.553	[illegible]	[illegible]	1	.603**	.628**
	Sig. (2-tailed)	<.00	[illegible]	[illegible]		<.001	<.001
	N	200	200	200	200	200	200
Trust	Pearson Correlation	.338**	.371**	.555**	.603**	1	.551**
	Sig. (2-tailed)	<.001	<.001	<.001	<.001		<.001
	N	200	200	200	200	200	200
Behavioural Intention	Pearson Correlation	.510**	.547**	.635**	.628**	.551**	1
	Sig. (2-tailed)	<.001	<.001	<.001	<.001	<.001	
	N	200	200	200	200	200	200

**. Correlation is significant at the 0.01 level (2-tailed).

Screen 8.18 Correlation Matrix

9 Factor analysis

Learning outcomes

At the end of this section, participants will be able to:

9.1 Recognise the purposes and requirements of factor analysis,
9.2 Enumerate the types of factor analysis,
9.3 Demonstrate factor analysis step by step using SPSS,
9.4 Interpret the factor analysis outputs, and
9.5 Report factor analysis in APA format.

9.1 The purposes and requirements of factor analysis

Factor analysis is used to simplify a complex variable by identifying the underlying or unobserved subfactors and also explaining the relationship between those subfactors. Ultimately, the outcome of a factor analysis is the identification of a grouping of correlated variables (i.e. a dimension, also interchangeably called a subfactor) from the complex variable (i.e. a construct, the explored factor).

A construct is a theory-based idea or concept; it is abstract and not directly observable. Construct helps us to understand or explain the phenomena, which is important for building theory and models. In contrast, dimension is a smaller measurable part of a construct. It is a specific aspect or component.

Factor analysis is part of the quantitative instrument development. It is also helpful to understand a theoretical concept, simplify data for large-scale information, and assess the construct validity of a scale. For example, a human resource manager wants to explore the organizational commitment of their employee. Organizational commitment is a construct and a broader concept and cannot be measured directly. Thus, the manager is interested in identifying the potential dimensions of organizational commitment. In fact, the dimensions consist of affective, continuance, and normative, which are useful to explain organizational commitment.

DOI: 10.4324/9781003673903-9

Requirements and assumptions of factor analysis

Scale of measurement

- A variable is measured using interval or ratio scales.

Sample size

- The ideal sample size may vary based on a few reasons, but the larger the sample size, the higher the accuracy. Nunally (1978) proposed a sample size of 10 times for each itemized scale. For example, a 10-item scale needs a minimum of 100 cases.
- Tabachnick and Fidel (2001) suggest a sample size of 300; however, the minimum sample of 150 is accepted if there are several high-loading variables (0.8 or more).

Assumptions

- The data should not have extreme values, that is, the absence of outliers.
- Meet the multivariate normality and have linear relationships between variables.
- No multicollinearity; extreme multicollinearity (high correlation between variables) or singularity (perfect correlation) should be avoided.

9.2 Types of factor analysis (FA)

There are two types of FA: exploratory and confirmatory. Although the selection of FA types is different, they have the same ultimate goal to find the underlying dimensions for the factor.

Exploratory factor analysis *(EFA)*

Select EFA if a study wants to develop a scale or instrument, specifically, the construct can be assessed directly. EFA is also used when there is limited knowledge about the relationship between the variables. A series of activities may take place before the researcher develops the scale, including literature review, focus group discussion, area experts meeting, and data collection.

Confirmatory factor analysis *(CFA)*

Proceed to this analysis if the researcher aims to confirm or disconfirm the extracted underlying dimensions or factor structure from a previous study (Tavakol & Wetzel, 2020). It can also be used when a researcher wants to validate the factor structure or dimensions of existing scales based on the theory, model, or previous study. Factor structure describes the relationships between variables; dimension describes the number of independent relationships (or factors) that exist.

Factor selection and rotation

Kaiser's criterion and the scree plot are two common methods used to determine the number of factors that exist.

Kaiser's criterion, a.k.a. Kaiser-Guttman rule or Kaiser-Guttman criterion (Kaiser, 1960), is a statistical method. Kaiser's criterion uses the eigenvalue rule and suggests retaining the item with an eigenvalue greater than 1.0. Eigenvalues refer to the amount of total variance explained by those factors. It is the sum of squared loadings for a factor (note that the factor here refers to dimension).

Scree plot, also known as Cattell's scree plot, is a graphical method introduced by Cattell (1966). The scree plot shows each eigenvalue of the factors, which starts steep and gradually flattens out. Refer to the scree plot; check a point where the curve changes direction and becomes horizontal. Retain all factors above the break and excluding the point at the break itself.

Next, **factor rotation** aims to simplify the complex factor loadings into a simpler factor structure without changing its underlying mathematical properties. It is a rotation process of transforming the factor axes in a multidimensional space, to find each variable loads strongly on only one factor and weakly on others. Two types of rotation that alter factor loadings and preserve the overall variance explained. **Orthogonal rotation** keeps the factors uncorrelated (independent). Choose orthogonal rotation when there is a theory, model, or empirical basis that assumes the underlying factors are independent. Examples of orthogonal rotation include *varimax*, *quartimax*, and *equamax*. Varimax is widely used because it maximizes the variance of the squared loadings within each factor and makes distinct concepts or dimensions. In contrast, **oblique rotation** (i.e. direct oblimin and promax) is selected when the factors allow for correlation with each other. Direct oblimin aims to simplify factors by minimizing the cross-products of loadings while simultaneously allowing for correlations between factors. Direct oblimin has a parameter (delta) that controls the degree of factor correlation. The default delta value is 0, which directs to a quartimin rotation, and the factors are not allowed to be correlated. **Promax** is a faster and computationally efficient method that combines a varimax rotation with a power transformation to simplify factors and allow for correlation. Promax uses a parameter (kappa) that influences the power to which factor loadings are raised, impacting the degree of simplification and correlation. In fact, the researcher can confirm which factor rotation to choose when they do their literature review.

Statistical measures

Bartlett's test of sphericity works to test whether the correlation matrix is significantly different from an identity matrix. The identity matrix is the opposite of a correlation matrix; it indicates no correlation between variables. If the test results are significant (p-value < significance level, often 0.05), it means the variables are correlated. In other words, the correlation matrix is not an identity matrix and is suitable for performing a factor analysis.

Kaiser-Meyer-Olkin (KMO) is a measure of sampling adequacy and a measure of the suitability of factor analysis. KMO value is a range between 0 and 1; the higher the better. Kaiser has categorized the acceptable KMO values as follows:

0.00 to 0.49 unacceptable
0.50 to 0.59 miserable
0.60 to 0.69 mediocre
0.70 to 0.79 middling
0.80 to 0.89 meritorious
0.90 to 1.00 marvellous

Field (2024) suggests that the KMO value of 0.7 or more reflects good internal consistency of the respondents.

Factor loadings

Factor loading is the correlation between items and the factor. The difference between an oblique and orthogonal rotation is that both the factor structure and factor pattern matrix are the same in orthogonal rotations. Whereas an oblique rotation produces both matrices. The correlations between the variables and the factors are called the factor structure matrix. Whereas the coefficients for the linear combination of the variables are presented in the factor pattern matrix. Note that Field (2024) recommended suppressing factor loadings less than 0.3 for EFA. This means that items below the factor loadings cut-off point of 0.3 will be removed in the analysis because it has a weak relationship with the underlying factor. Generally, factor loadings of 0.4 or higher are acceptable for social science studies.

Things to do after obtaining the extracted factors

- Rename the extracted factors because SPSS only provides numbers for the factors, such as factor 1 and factor 2.
- Run the reliability test for each factor. Check the potential items to be deleted from the scale, and refer to Cronbach's alpha if an item is deleted. Remember to use original items for positive statements and reverse-coded items for negative statements.
- Check for normality assumption for each factor.
- Run another reliability test if the extracted factors need to be combined as an index or composite scale. Check the reliability and the normality assumption for the combined factors.

Instrument and scales validation

Scale validation and replicability (e.g. split-half reliability) are very important stages of factor analysis in thesis writing. However, due to the scope and word limitations, the chapter is not able to provide detailed discussions.

Working example

The working data is part of a published paper by Yeoh et al. (2018) titled "Measurement Scale Development for Mobility-Related Quality of Life among Older Malaysian Drivers." The paper intended to develop mobility-related quality of life (M-QoL) scales based on the relevant theories and information from existing older drivers. It is an exploratory study; the area has limited knowledge and is unexplored, specifically in Malaysia, at the time when the study is ongoing. Hence, an exploratory factor analysis is suitable for this study instead of a confirmatory factor analysis, while the *promax rotation* under oblique rotation was chosen due to the researcher's understanding that the correlation between variables may exist after conducting a literature review.

9.3 Steps to perform a factor analysis

1. On the menu bar, select **File**, **Open**, and **Data** to open data (see Screen 9.1).

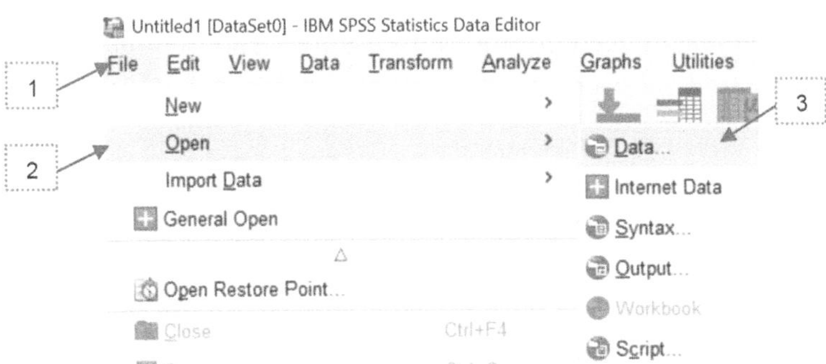

Screen 9.1 Data editor

2. In Screen 9.2, look in the sample data SPSS folder, select the **M-QoL** data file, and click **Open** to open the data file.

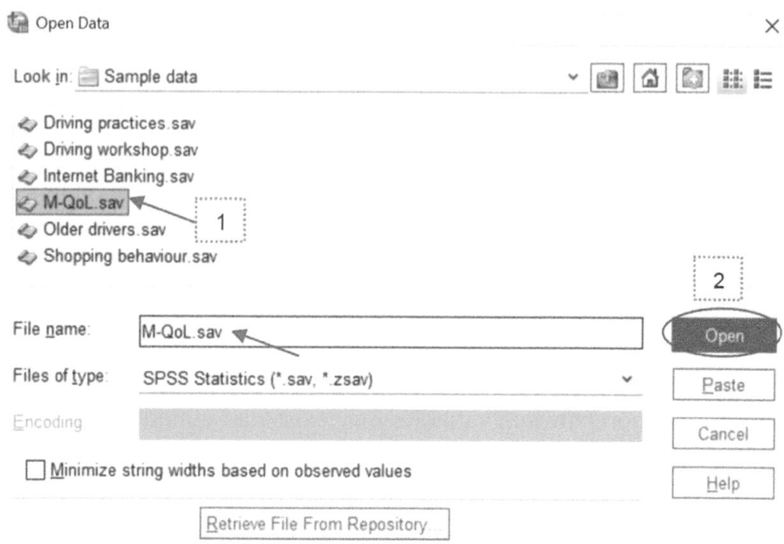

Screen 9.2 Open data

3. To choose the analysis, click **Analyze** in the main toolbar. This is followed by **Dimension Reduction,** and then click the **Factor** (see Screen 9.3).

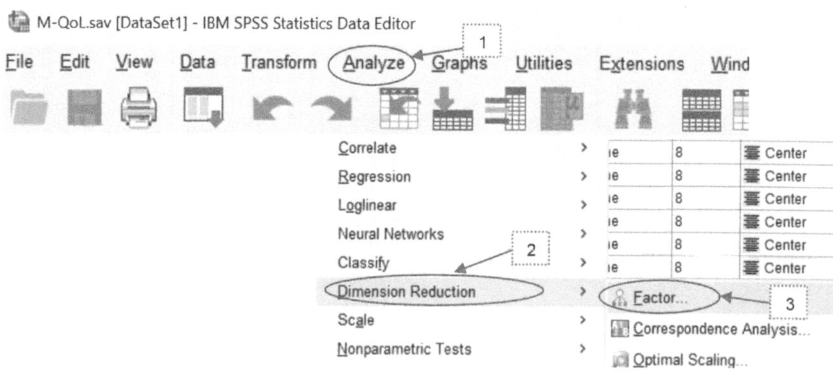

Screen 9.3 Choose analysis

4. In the variable list box, select the targeted items **g_1 to g_14** and move into **Variables** in Screen 9.4. (Beware, if there is a negative statement, insert the recoded items instead of the original negative items.)

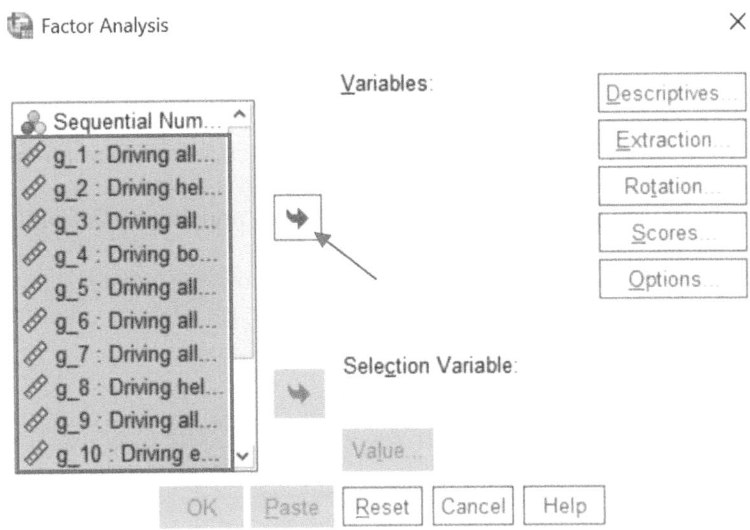

Screen 9.4 Factor analysis

5. Click on the **Descriptives**... and thus the dialog box **Factor analysis: Descriptives**. In Statistics, check ☑ **Univariate Descriptives**; under **Correlation Matrix**, check ☑ Coefficients, followed by ☑ KMO and Bartlett's test of sphericity, then click Continue in Screen 9.5.

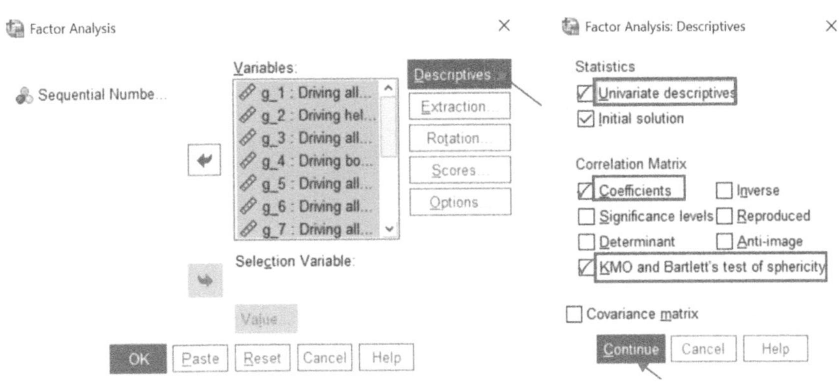

Screen 9.5 Factor analysis: Descriptives

6. Under **Extraction**, there are six extraction methods revealed; the default option is principal components analysis. The principal components analysis (PCA) is commonly selected by researchers; it aims to reduce the number of subfactors or variables while retaining most of the variance in the original variable. The second common method is maximum likelihood, which is used to determine the factor structure that best explains the observed data.

   In this working sample, the maximum likelihood has been chosen. In EFA, maximum likelihood is used to discover the factor structure that maximizes the likelihood of the observed correlation among the variables.

   Next, check ☑ the **scree plot** under **Display** and then click **Continue** in Screen 9.6.

   In general, follow the default setting of eigenvalues greater than 1 and maximum iterations for convergence 25. Note for CFA, an adjustment is needed for a fixed number of factors.

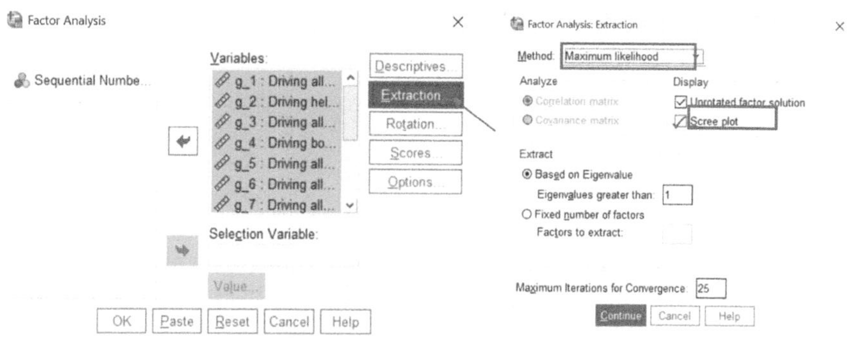

Screen 9.6 Factor analysis: Extraction

7. Under **Rotation**, choose Promax and Kappa and default at 4. Kappa is to control the degree of obliqueness; a higher delta value results in higher correlation between factors. Next, check ☑ the **Loading plot(s)** under **Display** and then click **Continue** in Screen 9.7.

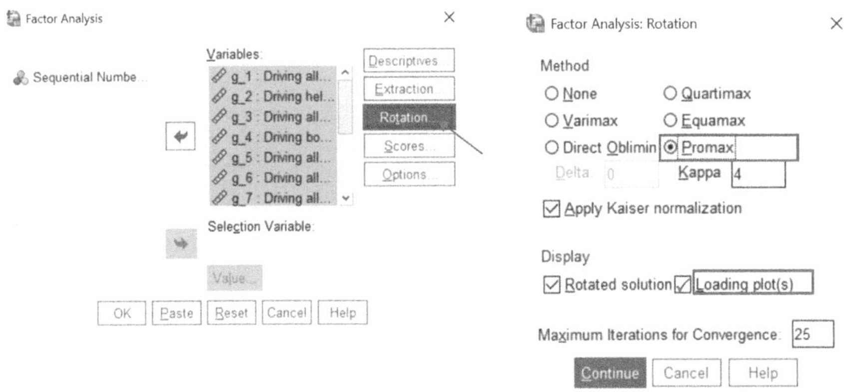

Screen 9.7 Factor analysis: Rotation

8. Under **Scores**, it is optional to save as variables. Generally, save a variable using the **Regression** method (see Screen 9.8). The factor scores can be used as a variable if the researcher wishes to perform subsequent analyses, such as multivariate analysis, ANOVA, and other multivariate analyses.

Screen 9.8 Factor analysis: Factor Scores

9. Under **Options**, missing values exclude cases listwise as the default setting. Then, under the coefficient display format, check ☑ **sorted by size** and ☑ **suppress small coefficients**, and adjust the absolute value to below 0 to 0.3, and click **Continue** in Screen 9.9. Note that items below the factor loadings cut-off point of 0.3 will be removed.

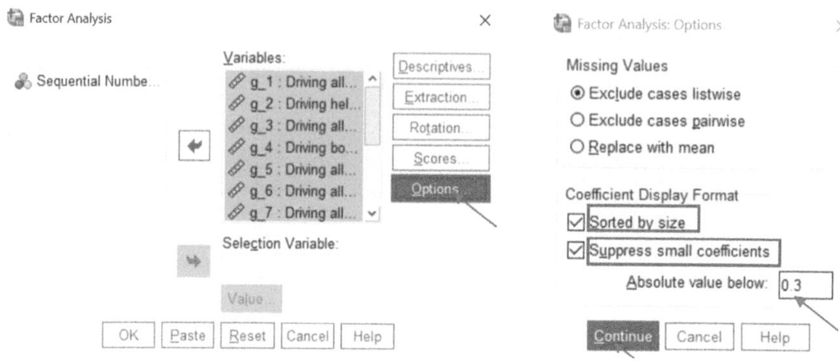

Screen 9.9 Factor analysis: Options

10. Lastly, click **OK** in Screen 9.10 to generate factor analysis outputs.

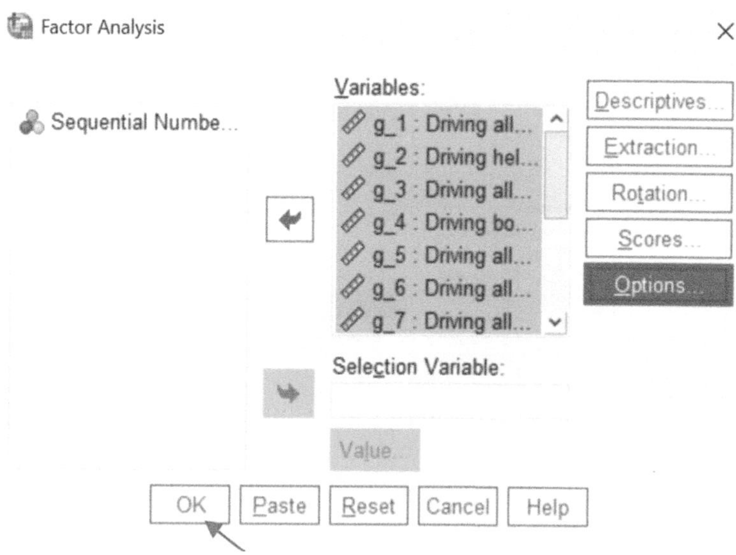

Screen 9.10 Run the factor analysis: Extraction

9.4 Hypothesis testing and interpretation of factor analysis SPSS outputs

1. State the null and alternative hypotheses for Barlett's test.

   Bartlett's test of sphericity is used to compare the observed correlation matrix from the data to an identity matrix.

   Null hypothesis

   H_o: The variables are uncorrelated (i.e. orthogonal, the correlation matrix is an identity matrix).

   Alternative hypothesis

   H_a: The variables are correlated (i.e. the correlation matrix is not an identity matrix).

Descriptive Statistics

	Mean	Std. Deviation	Analysis N
g_1 : Driving allows me to be mobile	4.09	.800	498
g_2 : Driving helps me to be independent	4.09	.752	498
g_3 : Driving allows me to enjoy things I like	4.01	.758	498
g_4 : Driving boosts my self esteem	3.75	.849	498
g_5 : Driving allows me to stay active	3.97	.749	498
g_6 : Driving allows me to do things I want whenever I want	4.11	.759	498
g_7 : Driving allows me to visit my family members	4.14	.720	498
g_8 : Driving helps me to stay connected with my friends	4.07	.713	498
g_9 : Driving allows me to participate in the community	3.92	.774	498
g_10 : Driving enables me to access bank services	3.85	.781	498
g_11 : Driving enables me to use post office services	3.79	.847	498
g_12 : Driving enables me to reach religious places	3.94	.848	498
g_13 : Driving enables me to go and take medication from clinic/ hospital	3.99	.769	498
g_14 : Driving enables me to go shopping complex	3.91	.825	498

KMO and Bartlett's Test

Kaiser-Meyer-Olkin Measure of Sampling Adequacy.		.906
Bartlett's Test of Sphericity	Approx. Chi-Square	3140.128
	df	91
	Sig.	<.001

Screen 9.11 Descriptive statistics and KMO and Bartlett's test

2. Based on the SPSS output in Screen 9.11, decide on the null hypothesis by using a 0.05 level of significance and conclude.

   Decision: Null hypothesis (H_0) is rejected, and therefore, H_a is supported.

   The corresponding p-value is lower ($p < .001$) than the .05 level of significance. A significant Bartlett's ($\chi 2 = 3140.13$, $p < .05$) test indicates that the observed correlation matrix is statistically different from the identity matrix. The variables are correlated; thus, applying a data reduction technique such as factor analysis is potentially useful for simplifying the data.
3. Based on the obtained Kaiser-Meyer-Olkin (KMO) in Screen 9.11, justify whether factor analysis is a suitable test.

   The KMO measure of sampling adequacy was 0.91, which is marvellous and above the acceptable value of 0.5. Thus, factor analysis is deemed to be a suitable test.
4. The communalities were referred to as the proportion of variance explained by the variable. Refer to the communalities output in Screen 9.12, and justify whether it is acceptable.

   Child (2006) suggests removing the variables with communality below 0.2. Generally, the communality between 0.4 and 0.6 is acceptable, 0.6 is considered ideal, while values below 0.4 are low. Note that this study has set the communality cut-off value at 0.4. Upon checking, only item g_9 has a lower communality value at 0.319; the item will be removed.
5. Refer to the Kaiser-Guttman criterion and scree plot (see Screens 9.13–9.14), and determine how many factors should be retained.

   Based on the number of eigenvalues 1 or above, it is recommended to retain two two-factor solutions. A total of 55.32% of the variance was

Communalities

	Initial	Extraction
g_1 : Driving allows me to be mobile	.437	.464
g_2 : Driving helps me to be independent	.463	.450
g_3 : Driving allows me to enjoy things I like	.518	.572
g_4 : Driving boosts my self esteem	.432	.419
g_5 : Driving allows me to stay active	.440	.427
g_6 : Driving allows me to do things I want whenever I want	.499	.551
g_7 : Driving allows me to visit my family members	.517	.447
g_8 : Driving helps me to stay connected with my friends	.519	.433
g_9 : Driving allows me to participate in the community	.319	.290
g_10 : Driving enables me to access bank services	.607	.674
g_11 : Driving enables me to use post office services	.615	.751
g_12 : Driving enables me to reach religious places	.420	.433
g_13 : Driving enables me to go and take medication from clinic/ hospital	.479	.454
g_14 : Driving enables me to go shopping complex	.426	.413

Extraction Method: Maximum Likelihood.

Screen 9.12 Communalities

explained by these two factors. Refer to the initial eigenvalues, the first factor contributed 43.62% of the variance, and the second factor accounted for 11.71% of the variance. Consistently, the scree plot shows a two-factor solution.

	Initial Eigenvalues			Extraction Sums of Squared Loadings			Rotation Sums of Squared Loadings[a]
Factor	Total	% of Variance	Cumulative %	Total	% of Variance	Cumulative %	Total
1	6.106	43.616	43.616	5.532	39.517	39.517	5.023
2	1.639	11.707	55.323	1.244	8.889	48.406	4.311
3	.998	7.131	62.454				
4	.721	5.149	67.604				
5	.698	4.985	72.589				
6	.643	4.589	77.178				
7	.544	3.885	81.063				
8	.483	3.447	84.510				
9	.435	3.107	87.618				
10	.402	2.872	90.489				
11	.394	2.812	93.302				
12	.362	2.585	95.886				
13	.346	2.473	98.359				
14	.230	1.641	100.000				

Extraction Method: Maximum Likelihood.

a. When factors are correlated, sums of squared loadings cannot be added to obtain a total variance.

Screen 9.13 Eigenvalues

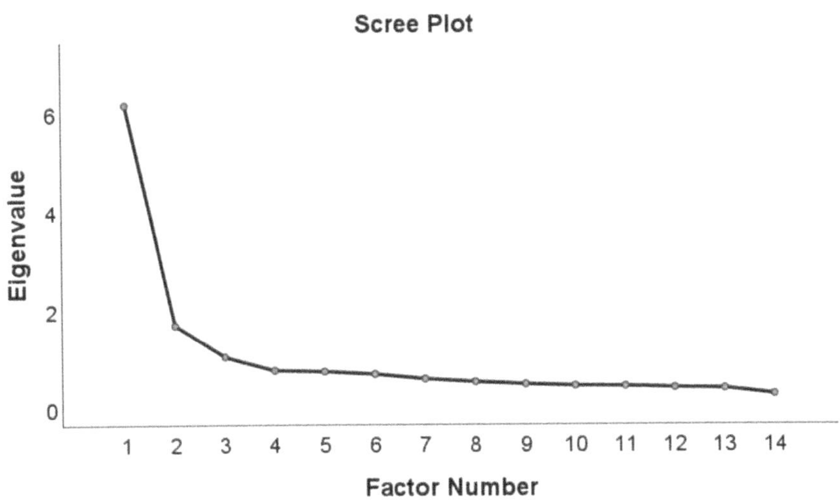

Screen 9.14 Scree plot

6. To check the consistency and stability of the factors derived from factor analysis. A reliability test using 13 items after excluding item g_9 was performed. Interpret the related Cronbach's alpha values (see Screen 9.15).

   All the statements have met the cut-off above 0.3 for the corrected item-total correlations. Further inspection of Cronbach's alpha value changes of the statement, if any, is deleted. Deletion of any statement does not increase the overall Cronbach's alpha value. This 13-item M-QoL scale, with a Cronbach's alpha of 0.895, is considered to have a good internal consistency (George & Mallery, 2024).

Reliability Statistics

Cronbach's Alpha	N of Items
.895	13

Item-Total Statistics

	Scale Mean if Item Deleted	Scale Variance if Item Deleted	Corrected Item-Total Correlation	Cronbach's Alpha if Item Deleted
g_1 : Driving allows me to be mobile	47.63	39.484	.583	.888
g_2 : Driving helps me to be independent	47.63	40.096	.559	.889
g_3 : Driving allows me to enjoy things I like	47.71	39.289	.644	.885
g_4 : Driving boosts my self esteem	47.97	39.098	.580	.888
g_5 : Driving allows me to stay active	47.75	39.936	.580	.888
g_6 : Driving allows me to do things I want whenever I want	47.61	39.527	.616	.886
g_7 : Driving allows me to visit my family members	47.58	39.798	.624	.886
g_8 : Driving helps me to stay connected with my friends	47.65	39.945	.613	.887
g_10 : Driving enables me to access bank services	47.87	39.297	.621	.886
g_11 : Driving enables me to use post office services	47.93	39.195	.572	.889
g_12 : Driving enables me to reach religious places	47.78	39.635	.528	.891
g_13 : Driving enables me to go and take medication from clinic/ hospital	47.73	39.486	.611	.887
g_14 : Driving enables me to go shopping complex	47.81	39.028	.609	.887

Scale Statistics

Mean	Variance	Std. Deviation	N of Items
51.72	45.984	6.781	13

Screen 9.15 Scree plot

7. Justify whether the EFA using oblique rotation is suitable for this study.

An intercorrelation analysis between the factors was performed using the regression factor scores (see Screen 9.16).

There is a significant positive relationship between the derived factors ($r = 0.653$, $p < 0.001$, $CI = 0.600–0.701$). This result can support the option of EFA using the oblique rotation, which indeed is appropriate in the study (Joseph et al., 2009).

Correlations

		REGR factor score 1 for analysis 1	REGR factor score 2 for analysis 1
REGR factor score 1 for analysis 1	Pearson Correlation	1	.653**
	Sig. (2-tailed)		<.001
	N	498	498
REGR factor score 2 for analysis 1	Pearson Correlation	.653**	1
	Sig. (2-tailed)	<.001	
	N	498	498

**. Correlation is significant at the 0.01 level (2-tailed).

Confidence Intervals

	Pearson Correlation	Sig. (2-tailed)	95% Confidence Intervals (2-tailed)[a]	
			Lower	Upper
REGR factor score 1 for analysis 1 - REGR factor score 2 for analysis 1	.653	<.001	.600	.701

a. Estimation is based on Fisher's r-to-z transformation.

Screen 9.16 Correlations and confidence intervals

8. Refer to Screen 9.17, and discuss the SPSS output of the goodness-of-fit test. Rename the subfactor based on the pattern matrix.

   The result of the goodness-of-fit test ($\chi^2 = 332.019$, $df = 64$, $p < 0.001$) is significant. This reflects that the observed data align well with the theoretical model that produced the pattern matrix (model validity). The produced pattern matrix works as evidence for the internal structure, in other words, construct validity. The constructs were measured effectively, and the relationships among items are consistent with the theoretical framework.

   Factor 1: Functions of driving
   Factor 2: Accessibility for independent living

Goodness-of-fit Test

Chi-Square	df	Sig.
332.019	64	.000

Pattern Matrix[a]

	Factor 1	Factor 2
g_6 : Driving allows me to do things I want whenever I want	.782	
g_3 : Driving allows me to enjoy things I like	.774	
g_2 : Driving helps me to be independent	.719	
g_1 : Driving allows me to be mobile	.701	
g_5 : Driving allows me to stay active	.637	
g_4 : Driving boosts my self esteem	.619	
g_7 : Driving allows me to visit my family members	.555	
g_8 : Driving helps me to stay connected with my friends	.468	
g_9 : Driving allows me to participate in the community		
g_11 : Driving enables me to use post office services		.957
g_10 : Driving enables me to access bank services		.836
g_12 : Driving enables me to reach religious places		.656
g_13 : Driving enables me to go and take medication from clinic/ hospital		.545
g_14 : Driving enables me to go shopping complex		.423

Extraction Method: Maximum Likelihood.
Rotation Method: Promax with Kaiser Normalization.
a. Rotation converged in 3 iterations.

Screen 9.17 Pattern Matrix

9.5 Report factor analysis in APA format

An exploratory factor analysis (EFA) using the promax oblique rotation was conducted to examine the underlying structure and the factorability of 14 items in M-QoL. The Kaiser-Meyer-Olkin measure of sampling adequacy was 0.91, which was above the cut-off value of 0.5. The Bartlett's test of sphericity was significant ($\chi^2 = 3140.13$, $p < 0.05$). This suggested that the item correlation matrix was significantly different from the identity matrix. The communalities were all above the cut-off of 0.4, further confirming that a common variance was shared among the items. Item 9, "driving allows me to participate in the community," with a factor loading of 0.34, was excluded.

Prepare a table of factor loadings using the pattern matrix (see Screen 9.17) and communalities (see Screen 9.12). Table 9.1 is the sample template; see the complete example table in Yeoh et al. (2018).

The factor extraction was consistent with the questionnaire development (i.e. two factors), and the items also loaded consistently with the earlier assumption, which was made based on the literature review and the focus group discussion. The first eight items reflect the functions of driving, which contribute to the quality of life, including the drive to participate in social activities (Gabriel & Bowling, 2004). Items 9 to 14 were related to driving as a mode to assist individuals' daily functioning through enabling access to facilities and achieving independent living (Graf, 2009).

The maximum likelihood analysis was used to identify and compute the composite M-QoL scores for the underlying factors. The eigenvalues and the percentage of variance are presented in Table 9.2. In terms of eigenvalues that are equal to 1.0 or higher, a two-factor solution was found. The two factors account for 55.32% of the variance in total. The initial eigenvalues show that the first factor explained 43.62%, and the second factor explained 11.71% of the variance.

Table 9.1 Factor loadings (pattern matrix) and communalities of M-QoL scale

Item no.	*Statements*	*Factor loadings*	*Communalities*
	Factor 1: Functions of driving		
1.	Driving allows me to be mobile	0.70	0.46
	Cont. item 2 to 8		
	Factor 2: Independent living/ accessibility		
10	Driving enables me to access bank services	0.84	0.67
	Cont. item 11 to 14		

Extraction method: maximum likelihood; rotation method: promax with Kaiser normalization; factor loadings less than 0.4 have not been presented.

Note: Adapted from the reliability test by Yeoh et al. (2018), copyright 2018 by Springer.

Table 9.2 Eigenvalues and percentage of variance for M-QoL from EFA

Factor	*Eigenvalue*	*Percentage of Variance*	*Cumulative Percentage of Variance*
1	6.11	43.62	43.62
2	1.64	11.71	55.32

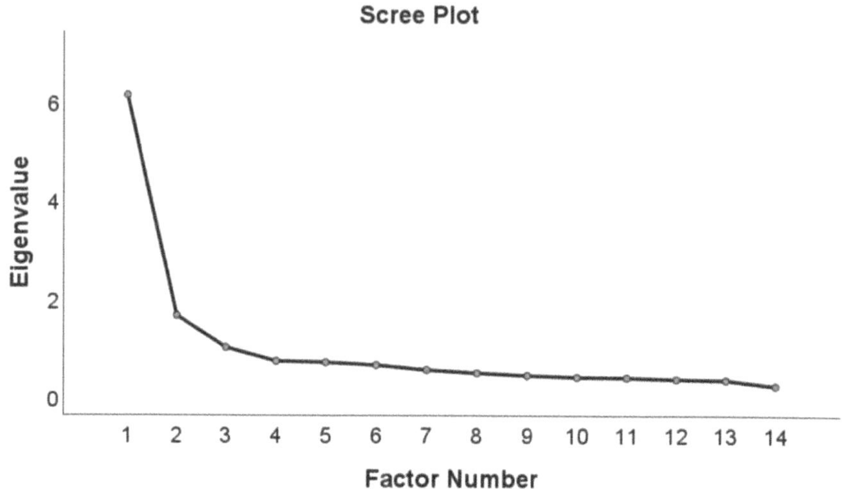

Screen 9.18 Scree plot of the maximum likelihood analysis of the M-QoL scale

The scree plot analysis also suggested a two-factor solution (see Screen 9.18). Consistently, the two-factor structure was suggested for this scale development. Factor 1 was renamed as functions of driving, whereas Factor 2 was labelled as independent living or accessibility.

The reliability test was performed to check the internal consistency. The Cronbach's alpha of the M-QoL scale is 0.90, which indicates the scale has a good internal consistency (George & Mallery, 2024). All items had the corrected item-total correlations above 0.3. Besides that, deletion of any item does not increase the Cronbach's alpha value. Thus, it is recommended that all 13 items be retained in the M-QoL scale.

Prepare a table for the reliability test. Table 9.3 is the template. Complete the statistics based on Screen 9.15.

Table 9.3 Reliability analysis of 13-item M-QoL scale

		Scale mean if item deleted	*Scale variance if item deleted*	*Corrected item-total correlation*	*Cronbach's alpha if item deleted*
1	Driving allows me to be mobile	47.63	39.48	0.58	0.888
	Cont. item 2 to 13 (remember that item 9 was excluded due to low communalities, see Screen 9.12)				
14	Driving enables me to go shopping	47.81	39.03	0.61	0.887

Cronbach's alpha = 0.90; *Var* = 45.98; *SD* = 6.78; number of items = 13

Note: Adapted from the reliability test by Yeoh et al. (2018), copyright 2018 by Springer.

Table 9.4 Psychometric properties of the 13-item M-QoL scale and its subscales

Items	*No. of items*	M	SD	*Skewness*	*Kurtosis*	*Cronbach's alpha* (α)
M-QoL	13	51.72	6.781	0.07	−0.38	0.90
Functions of driving	8	32.24	4.43	−0.04	−0.36	0.87
Accessibility for independent living	5	19.48	3.20	−0.10	−0.14	0.84

Reliability testing for the subscale

The reliability test results of M-QoL and its subscales are summarised in Table 9.4. The Cronbach's alpha values for the subscales are 0.84 and 0.87, respectively. A close inspection of any item elimination and the changes in alpha values was done. Deletion of any of the items did not increase the alpha values. The value of corrected item-total correlation was above 0.3 for both subscales.

The composite scores were generated for M-QoL for hypothesis testing purposes. Higher scores indicated higher mobility-related quality of life achieved by older drivers. Based on skewness (−0.10 to 0.07) and kurtosis (−0.14 to −0.38), the data of M-QoL includes two subscales that were about normally distributed. Since the assumption of a normal distribution is met, the parametric statistical analysis can be used to test the hypothesis (George & Mallery, 2024). In addition, there was a significant and positive correlation between the functions of driving and accessibility ($r = 0.653$, $p < 0.05$). This reaffirmed the appropriateness of using the oblique rotation in EFA.

10 Multiple linear regression

Learning outcomes

At the end of this section, participants will be able to:

10.1 Recognise the purposes and requirements of multiple linear regression (MLR) analysis,
10.2 Demonstrate MLR step by step using SPSS,
10.3 Interpret the MLR outputs, and
10.4 Report MLR in APA format.

10.1 Purpose and requirements of multiple linear regression analysis

MLR analysis is considered an advanced level of quantitative analysis. It is worth introducing in this book because it is an extension of simple linear regression from the beginner level. Both simple linear regression and MLR are used to understand the relationship between the independent (X) variables and a dependent variable (Y), and how these X collectively impact Y. Simple linear regression predicts the dependent (outcome) value involves only one independent variable (predictor) and determines how well the model fits the overall data. MLR includes more than one independent variable and determines which independent variable is a significant predictor.

The scale of measurement for both testing variables (X and Y) requires continuous data (i.e. either interval or ratio), or a dummy variable is acceptable for the independent variable. To obtain a valid MLR analysis, there are five assumptions to be fulfilled. Those are normality, linearity, homoscedasticity, independence, and no multicollinearity. Any violation of the assumptions may cause the MLR results to be unreliable.

1. Linearity: An approximate linear relationship exists between the predictor and outcome variable. A scatterplot is used to visualize the relationship between the predictor and each outcome variable, the potential outlier, and heteroscedasticity.

DOI: 10.4324/9781003673903-10

2. Independence: In the data set, the observations (subjects) are independent from each other and not influenced by other observations.
3. Normality: The residuals (the differences between the predicted and actual values) of the model are approximately normally distributed. Checking this assumption uses (i) graphical method – histogram/normal Q-Q plot or (ii) statistical method (normality test refers to the Kolmogorov-Smirnov or Shapiro-Wilk.
4. Homoscedasticity: The constant variance of the residuals (errors) exists across the independent variables. Residuals or errors are the difference between the observed values and the predicted values from the obtained regression model. Refer to the scatterplot; homoscedasticity refers to the points spread consistently across the range of predicted values. If there is a violation (heteroscedasticity), the points spread wider at predicted values, and the patterns look like a cone shape.
5. No multicollinearity: Absence of high correlation between the predictor variables. Multicollinearity happens when one or more predictor variables are highly correlated, making the coefficient estimates unreliable or difficult to interpret the unique effect of the predictors. If the tolerance value is more than 0.10 and the VIF statistics are less than 10.0, it is indicative of no multicollinearity.

Components of MLR

The outcome of MLR is to find the prediction model and the regression equation produced based on the collected observations.

$$Y = \beta_0 + \beta_1 X_1 + \beta_2 X_2 + \ldots + \varepsilon$$

1. Independent Variable (X_1, X_2, . . .): The predictors, causes, explanatory variables, or inputs that reasonably influence the dependent variable.
2. Dependent Variable (Y): The outcome variable, effect, or output that you predict using the obtained regression model.
3. Coefficients (β_0, β_1, β_2, . . .): The change in Y when there is a one-unit change in the corresponding X, holding all other Xs are constant.

   - **β_0 (Intercept):**

   It is constant and represents the predicted value of Y, and all Xs are 0.

   - **β_1, β_2, . . . , β_β (Slopes):**

   Slopes reflect the strength and direction of the relationship between each X and Y.
4. Residuals/Error (ε): The unexplained variation in the Y, the difference between the observed values and the predicted values of Y from the obtained regression model.

Besides the equation, the R-square (a.k.a. coefficient of determination) denotes the proportion of variance in Y that is explained by the selected X. R-square value ranges between 0 and 1, which can be converted to a percentage from 0% to 100%. The R-square value close to 1 show that a better model fits and the model explains a larger proportion of the variance in the Y.

Preparation before MLR SPSS procedure

Assuming that data cleaning, such as missing value and outlier detection, and also the reliability test and normality test, has been done. Do the data transformation for the multi-item scale to tabulate the summation score, and always be mindful of negative statements. To get the data ready, check if you have a categorical variable in your independent variable. If a categorical variable exists, create a dummy variable using data transformation to avoid errors.

For example, gender is categorical data. Select one category as the reference category and label it as 1; the rest will be 0. Transform and recode them, then save a new variable. Later on, insert the newly created dummy variable into the MLR analysis instead of the original coding.

Variable name: Gender	*Variable name: Dummy_Gender*
Original coding 1: Female 2: Male	Transform to 0: Female 1: Male

If the variable is multi-categorical data (more than two categories). Decide the number of dummy variables by rule k-1, assuming that k is the number of categories in the particular variable. Let's say there are three levels of education, k = 3 – 1 = 2, thus creating two dummy variables. Decide on the reference category; if primary is selected as the reference category, label it as 0.

Variable name: Educational_level	*Variable name: Dummy1_Secondary*
Original coding 1: Primary 2: Secondary 3: Tertiary	Transform to 0: Primary 1: Secondary 0: Tertiary
	Variable name: Dummy2_Tertiary
	Transform to 0: Primary 0: Secondary 1: Tertiary

Be cautious, only enter two dummy variables and exclude the reference group to avoid the dummy variable trap or multicollinearity.

Working example

The working data is part of a study from Yeoh et al. (2009).

Research objective

To predict the impact of driving knowledge (X_1), driving attitude (X_2), and driving confidence (X_3) on driving practices (Y) among older Malaysian drivers.

Independent variable: X_1, X_2, X_3
Dependent variable: Y

* *All variables were collected using multiple items. Before the MLR analysis, do preliminary data checking for outliers and missing data, as well as reliability for the scales. Most importantly, the data computation needs to be done for the summation score and data transformation for the categorical data, if there is any, for the independent variables.*

Perform a correlation matrix (refer to Chapter 8) to check if the variables are significant, no singularity (perfect zero correlation), or perfect correlation ($r > 0.8$) with the independent variables.

It is meaningless to include an insignificant variable. A perfect correlation indicates the potential condition for multicollinearity.

10.2 Multiple linear regression procedure using SPSS

Refer to the sample research objective in Chapter 10.1. This section provides the SPSS process on MLR using the enter method.

1. On the menu bar, select **File**, **Open**, and **Data** (see Screen 10.1).

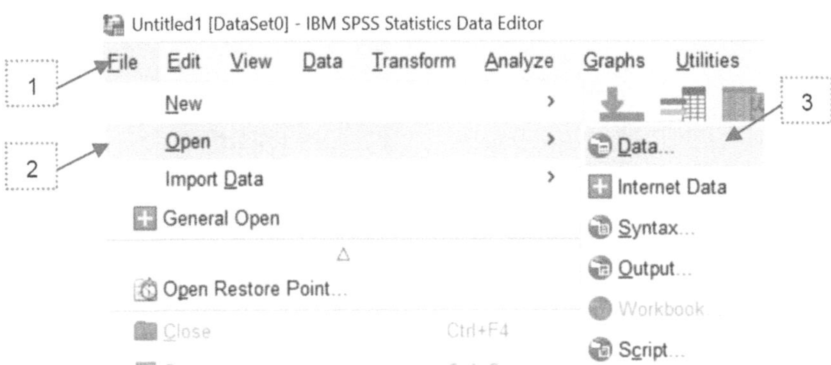

Screen 10.1 Data editor

2. In the sample data SPSS folder, select the **Driving Practices** data file and click **Open** to open the data file (see Screen 10.2).

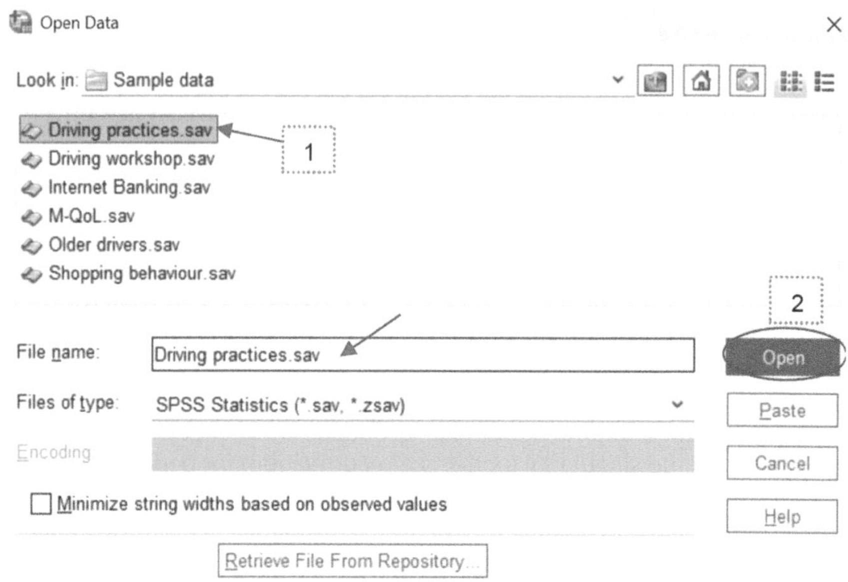

Screen 10.2 Open data

3. Go to the main menu, click **Analyze**, followed by **Regression**, and select **Linear** (see Screen 10.3).

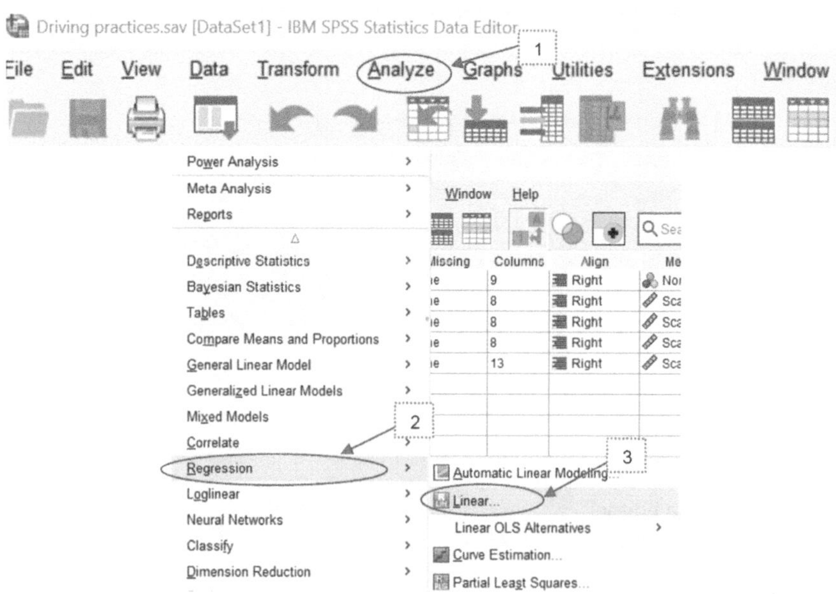

Screen 10.3 Choose analysis

4. In Screen 10.4, select the **Sum of Driving Practices**, move into the **Dependent**. Next, select the independent variable (**Sum of driving knowledge, driving attitude, driving confidence**) and move it to the **Independent(s)**. Use the default **Enter** method to enter all variables and regress all of them simultaneously.

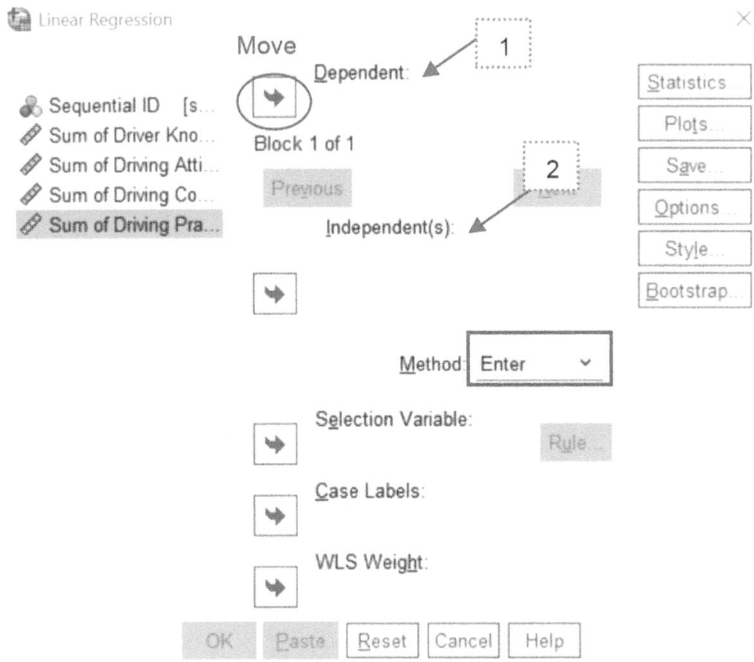

Screen 10.4 Linear Regression

There are **five** (5) types of methods under linear regression:

Enter: This option is entering all the selected independent variables and performing regression analysis at one time, regardless of their significance.

Stepwise: This is a mixed method of **Forward** and **Backward**. It is useful to test the effect of an additional variable, especially when the additional variable changes the variance of the model or affects the predictive validity of other variables. This method will remove the variable that lowers the variance of the predictive model.

Remove: This method starts with the **Enter** method and includes all the independent variables in Block 1 of 1. Then, manually remove one variable each time in the next block (Block 2 of 2, Block 3 of 3, etc.). Interpret the model that best explains the variance of the dependent variable.

Backward: This method also starts with the **Enter** method, and the system will automatically remove the most insignificant variable one at a time. The process repeats until it excludes all the insignificant variables and retains all the significant variables.

Forward: This method is the opposite of the **Backward** method. The most significant variable will be added at each step. Repeat this process and stop when adding more variables does not explain additional variance.

5. In Screen 10.5, **Linear Regression: Statistics**, under the **Regression Coefficients**, check the option **Confidence intervals at 95% and Descriptives**.

   Secondly, check the **Durbin-Watson** under **Residuals** and click **Continue**.

   *Check the R-square change if you want to test the weightage of the additional variable. Besides, if the assumption testing of autocorrelation, collinearity, and outliers has not been carried out yet, you may check the additional options, including part and partial correlations, collinearity, and case-wise diagnostics.

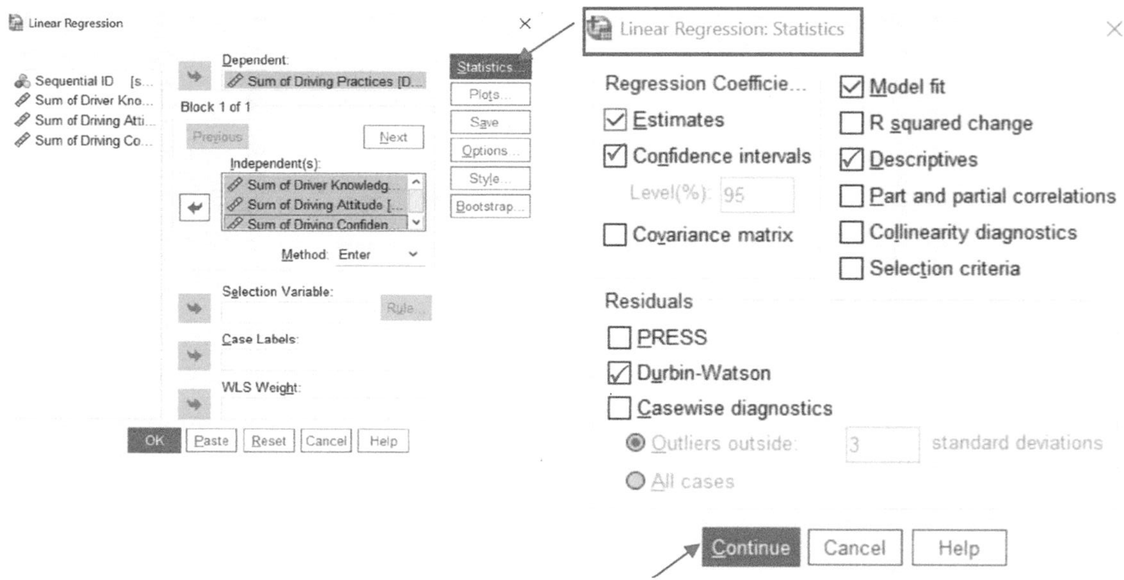

Screen 10.5 Linear Regression: Statistics

6. In Screen 10.6 **Linear Regression: Plots**, move *ZPRED to Y and *ZRESID to X to create a scatter plot. Check both options of the **Standardized Residual Plots** and click **Continue**.

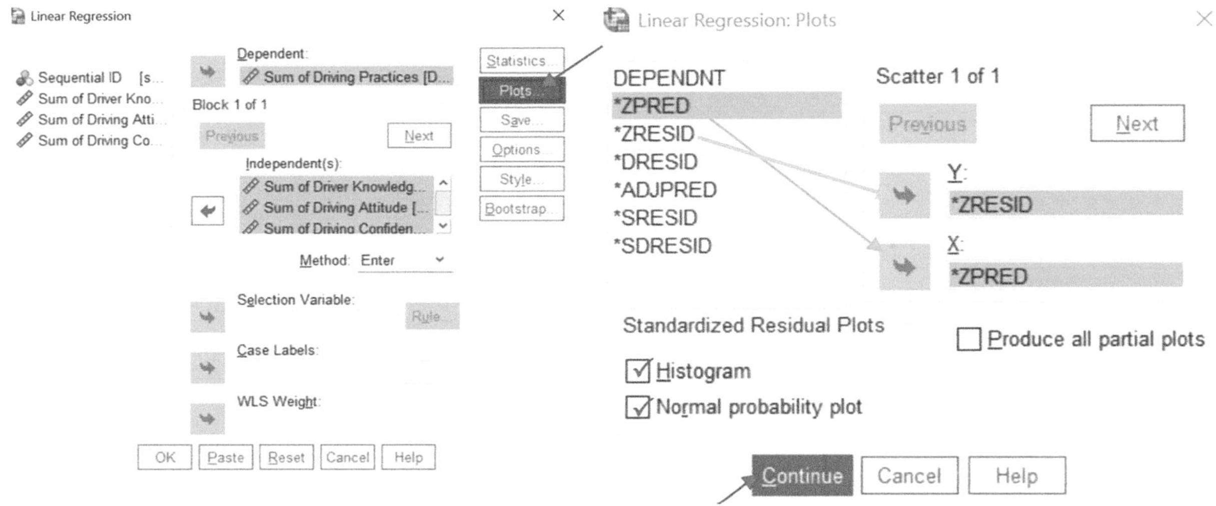

Screen 10.6 Linear Regression: Plots

7. The function of **Linear Regression: Save** is to save as a new variable for the selected options for further analytical purposes. In this sample, we check the options of **Unstandardized** under **Predicted Values** to see the raw predicted values of dependent variables. Then, click **Continue** and lastly **OK** to run this MLR analysis (Screens 10.7–10.8).

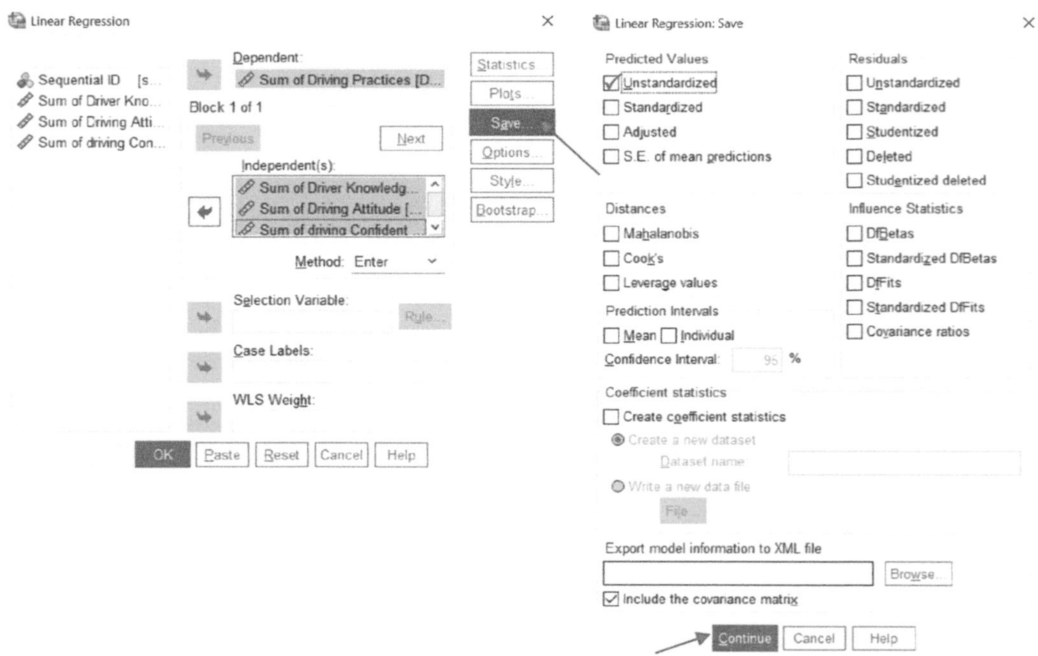

Screen 10.7 Linear Regression: Save

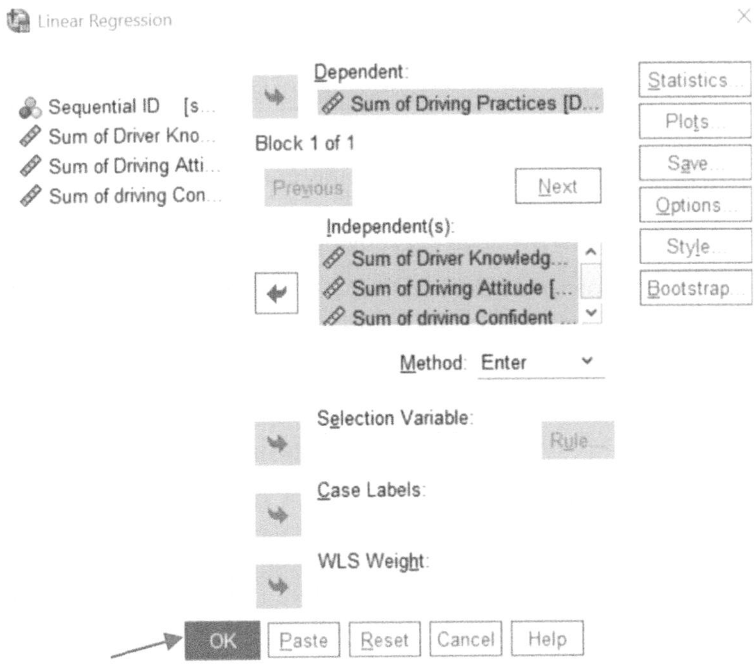

Screen 10.8 Run the analysis

Look at the variable view, a new variable, **PRE_1**, was created to show the unstandardized predicted values (see Screen 10.9).

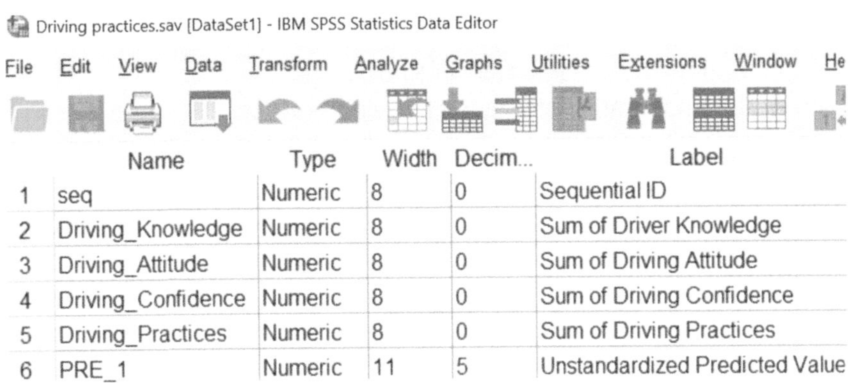

	Name	Type	Width	Decim...	Label
1	seq	Numeric	8	0	Sequential ID
2	Driving_Knowledge	Numeric	8	0	Sum of Driver Knowledge
3	Driving_Attitude	Numeric	8	0	Sum of Driving Attitude
4	Driving_Confidence	Numeric	8	0	Sum of Driving Confidence
5	Driving_Practices	Numeric	8	0	Sum of Driving Practices
6	PRE_1	Numeric	11	5	Unstandardized Predicted Value

Screen 10.9 Variable view

10.3 Selected output of MLR analysis and data interpretation

1. State the null and alternative hypotheses based on the research objective in Chapter 10.1.

   Ho: Variance in driving practices of older Malaysian drivers will not be explained by driver knowledge, driving attitude, and driving confidence.
   Ha: Variance in driving practices of older Malaysian drivers will be explained by driver knowledge, driving attitude, and driving confidence.

2. What can you interpret based on the *Model summary* in Screen 10.10?
   The obtained R-squared is 0.120, which indicates that 12% of the driving practices (dependent variable) were explained by the linear combination of the significant independent variables.

Descriptive Statistics

	Mean	Std. Deviation	N
Sum of Driving Practices	34.80	3.400	398
Sum of Driver Knowledge	8.49	1.316	398
Sum of Driving Attitude	22.11	2.592	398
Sum of Driving Confidence	16.70	3.487	398

Model Summary[b]

Model	R	R Square	Adjusted R Square	Std. Error of the Estimate
1	.346[a]	.120	.113	3.202

a. Predictors: (Constant), Sum of Driving Confidence, Sum of Driving Attitude, Sum of Driver Knowledge
b. Dependent Variable: Sum of Driving Practices

Screen 10.10 Descriptive statistics and model summary

3. Refer to the ANOVA in Screen 10.11. Is the proposed model significant? Justify your answer based on a 0.05 level of significance.

ANOVA[a]

Model		Sum of Squares	df	Mean Square	F	Sig.
1	Regression	551.006	3	183.669	17.914	<.001[b]
	Residual	4039.509	394	10.253		
	Total	4590.515	397			

a. Dependent Variable: Sum of Driving Practices
b. Predictors: (Constant), Sum of Driving Confidence, Sum of Driving Attitude, Sum of Driver Knowledge

Screen 10.11 ANOVA

The obtained p-value (<.001) is smaller than the .05 level of significance. This showed that the overall model is significant, F (3, 397) = 17.914, p < .001, and the independent variables explained a significant amount of variance.

4. Present the prediction equation and check the collinearity statistics.

   Refer to Screen 10.12, all three predictor variables (driver knowledge, driving attitude, and confidence) were significant ($p < .05$).

   Refer to the unstandardized coefficients, the equation is as follows.

   Driving practices = 22.60 + 0.31 driving knowledge + 0.30 driving attitude + 0.18 driving confidence

   The constant is positive, which reflects that the regression slope is positive.

   The obtained collinearity tolerance value, ranging between 0.97 and 0.98, shows the absence of collinearity because the rule of thumb of no collinearity is a tolerance value of more than 0.10, while the cut-off VIF (VIF= 1/Tolerance) statistics is very close to 0 or less than 10 (Field, 2024). Since the result indicates the VIF ranges between 1.01 and 1.03, collinearity is not found. Consistently, a condition index of less than 30 is an indicator of free collinearity, and the obtained condition index is 25.96 (see Screen 10.13).

5. Report the coefficient of determination and interpret its value.

   Refer to the standardized coefficient because the variables have different measurement units to compare. All three predictors are significantly (p < 0.05) influenced by driving practices. The standardized coefficient value is the highest for driving attitude (β = 0.23), followed by driving confidence (β = 0.18) and driving knowledge (β = 0.12). The direction of all three predictors has a positive influence on the driving practices (all beta values are positive).

6. Visualize the obtained histogram, normal P-P plots of regression standardized residuals, and scatterplot, and describe them (see Screens 10.14–10.16).

Both the histogram and the normal P-P plot of regression standardized residuals are used to check the normality of residuals. The normal P-P plot shows that all observed values fall roughly along the theoretical normality line, which means that residuals were normally distributed. The scatterplot illustrated homoscedasticity because the points spread consistently across the range of predicted values, and the residual variances were about equal or constant.

Coefficients[a]

Model		Unstandardized Coefficients		Standardized Coefficients			95.0% Confidence Interval for B		Collinearity Statistics	
		B	Std. Error	Beta	t	Sig.	Lower Bound	Upper Bound	Tolerance	VIF
1	(Constant)	22.595	1.728		13.073	<.001	19.197	25.993		
	Sum of Driver Knowledge	.308	.124	.119	2.482	.013	.064	.552	.968	1.033
	Sum of Driving Attitude	.300	.063	.229	4.797	<.001	.177	.423	.984	1.017
	Sum of Driving Confidence	.177	.047	.182	3.775	<.001	.085	.269	.966	1.036

a. Dependent Variable: Sum of Driving Practices

Screen 10.12 Coefficients

Collinearity Diagnostics[a]

Model	Dimension	Eigenvalue	Condition Index	Variance Proportions (Constant)	Sum of Driver Knowledge	Sum of Driving Attitude	Sum of Driving Confidence
1	1	3.945	1.000	.00	.00	.00	.00
	2	.031	11.252	.01	.09	.04	.96
	3	.018	14.994	.02	.76	.27	.01
	4	.006	25.957	.96	.15	.69	.03

a. Dependent Variable: Sum of Driving Practices

Screen 10.13 Collinearity diagnostics

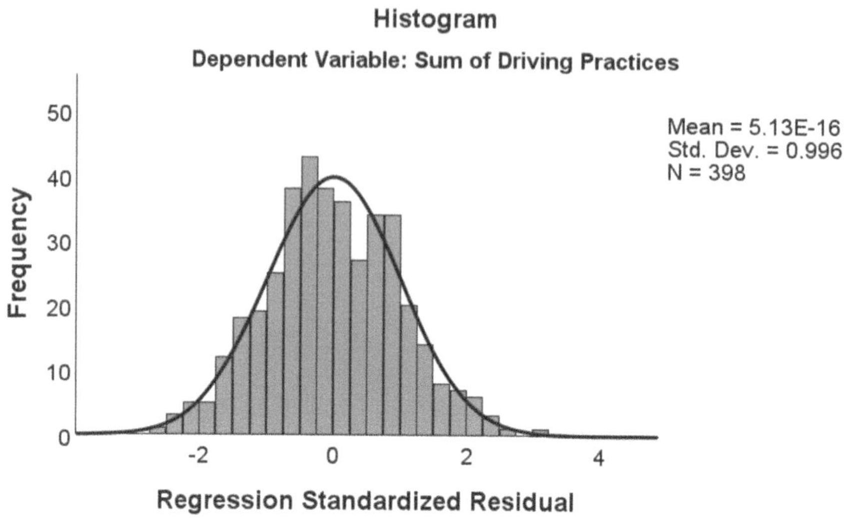

Screen 10.14 Histogram of regression standardized residual

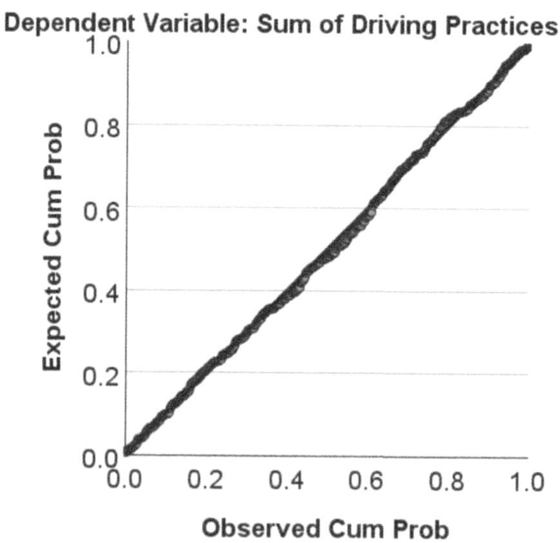

Screen 10.15 Normal P-P plot of regression standardized residual

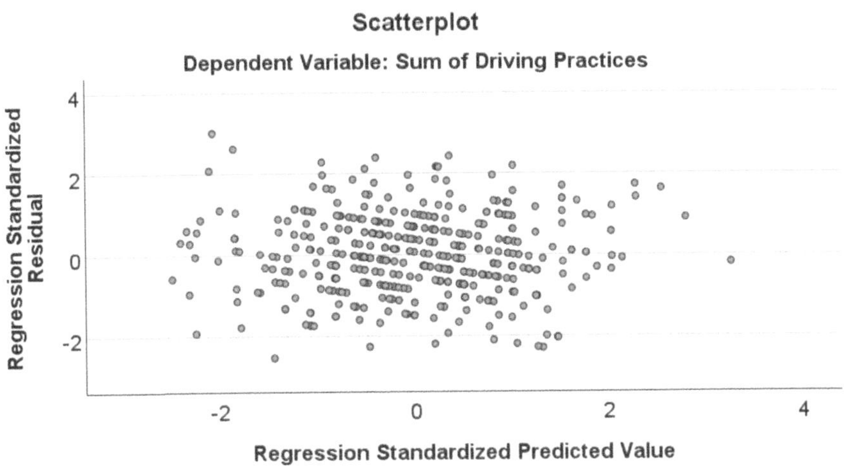

Screen 10.16 The scatterplot of standardized predicted values versus observed values

10.4 Sample write-up of MLR in APA format

Descriptive analysis for the variables

Generally, older Malaysian drivers in this study reported practicing good driving (M = 34.80, SD = 3.40). They have a high score in driving knowledge (M = 8.49, SD = 1.32) and have a relatively good attitude towards driving (M = 22.11, SD = 2.59). The older drivers showed confidence when driving in specific situations (M = 16.70, SD = 3.49).

Results and brief interpretation of the MLR analysis

All the selected independent variables have a significant relationship at the bivariate level (Correlation) with a weak strength. Table 10.1 summarizes the multiple linear regression analyses in driving practices.

The F-statistic for the overall goodness of fit of the model is 17.91 ($p < 0.001$). The following regression model was produced using the enter method.

Driving practices = 22.60 + 0.31 driving knowledge + 0.30 driving attitude + 0.18 driving confidence

This model is considered a moderate descriptor because the model only explains 12% of the variance in driving practices [F (3, 394) = 17.91, $p <$ 0.001]. The R-squared value of 0.12 indicated that 12% of the driving practices (Y) were explained by the linear combination of the three-predictor

Table 10.1 Summary of multiple linear regression analyses for driving practices

Variables	*Unstandardized coefficients*		*Beta (standardized coefficient)*	*t*
	B	*SE (standard error)*		
(Constant)	22.60	1.73		13.07**
Sum of Driver Knowledge	.308	0.12	0.12	2.48*
Sum of Driving Attitude	.300	0.06	0.23	4.80**
Sum of Driving Confidence	.177	0.05	0.18	3.78**
F (3, 394) = 17.91, $p < 0.001$, R = 0.35, R square = 0.12, Adjusted R square = 0.11				

* Significant at the 0.05 level of significance
** Significant at the 0.01 level of significance

variables. Given that, the obtained p-value was smaller than the significance level at 0.05. Thus, it is statistically evident that the slope of the estimated linear regression model is not equal to 0. This further confirms that the research data are fit to the three anticipated predictors (X) in this study.

Close inspection of the three predictors, driving attitude contributes most to the driving practices of the respondent ($\beta = 0.23$). Then followed by driving confidence ($\beta = 0.18$) and driving knowledge ($\beta = 0.12$). The value of the standardized beta coefficient is used to estimate the changes of X and Y. This means that, when one standard deviation increases in driving attitude score (X), there is a 0.23 standard deviation increase in the driving practices (Y). It reflects that those drivers have a positive driving attitude, higher knowledge, and high confidence when practicing safe driving. To conclude, Ho is rejected and Ha is accepted.

According to the obtained collinearity diagnostic table, none of the tolerance values is smaller than 0.10 (0.97 to 0.98), and the VIF statistics are less than 10.0 (1.01 to 1.03). This showed that there is no multicollinearity problem among the predictor variables of the estimated model (Field, 2024). Consistently, a condition index of less than 30 is an indicator of free collinearity (condition index is 25.96). The normal P-P plot of regression standardized residuals, as in Figure 10.1, revealed that all observed values fall roughly along the theoretical normality line, indicating that the residuals were from a normally distributed population.

The scatterplot as depicted in Figure 10.2 indicates that the relationship between the dependent variables and the predictors was linear and the residual variances were about equal or constant.

Note that there is no multicollinearity problem between the predictors included in the model. Apart from that, the assumptions of MLR include normality, equality of variance, and linearity, which are also all met. Thus, it is logical to conclude that this estimated MLR model is acceptable to explain

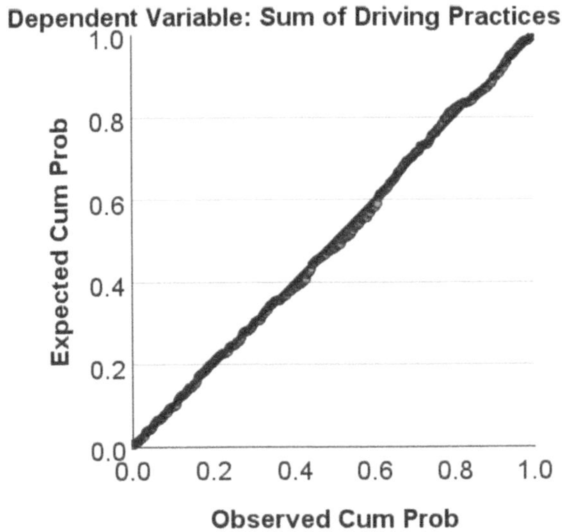

Figure 10.1 Normal P-P plot of regression standardized residual

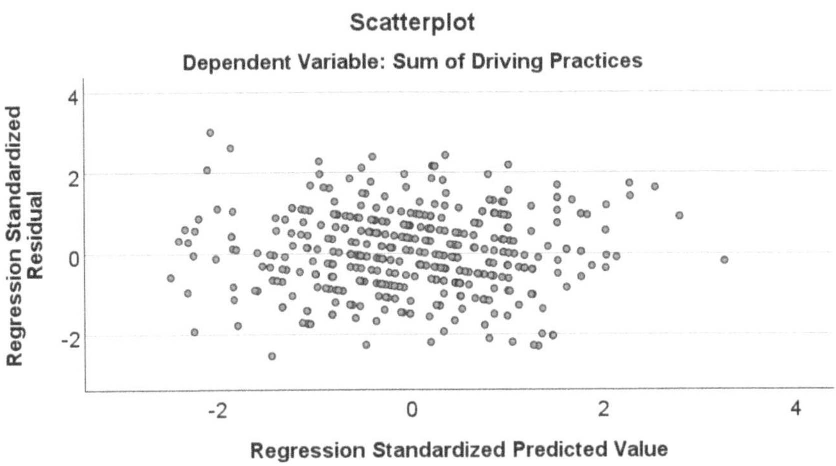

Figure 10.2 The scatterplot of standardized predicted values versus observed values

the variation of driving practices. Although it is a moderate descriptor, it is a stable and valid model.

** *Lastly, insert discussion and results comparison with past research studies.*

11 Conclusion

Learning outcomes

At the end of this section, participants will be able to know:

11.1 The difference between descriptive analysis and hypothesis testing,
11.2 The criteria to choose an appropriate analysis method, and
11.3 An example of research objectives and types of analysis applied.

11.1 The difference between descriptive analysis and hypothesis testing

What is descriptive analysis?

It helps in describing, showing, or summarizing data in a meaningful way. These include the statistics of:

- Frequency, percentage
- Measures of central tendency (mode, median, and mean)
- Measures of spread (range, quartiles, absolute deviation, variance, and standard deviation)

For data presentation, these data can be displayed in a graphical method; see Table 11.1. Figure 11.1 shows an example of a pie chart by percentage.

What is hypothesis testing?

It is a statistical method that is used in making statistical decisions using experimental/survey data about a sample or population.

11.2 The criteria to choose an appropriate analysis method

How do we identify types of quantitative analysis?

Based on your:

1. Research questions/objective/purpose of study
2. Questionnaire design/scale of measurement
3. Assumption testing (e.g. normality test, equality of variance)

DOI: 10.4324/9781003673903-11

Table 11.1 Graphical options for difference statistics

Types of descriptive statistics	*Graphical options*
Frequency and percentage	Histogram, pie charts, bar charts
Measures of central tendency	Histogram, boxplot, specialized graphs
Measure of spread	Box plot, histogram

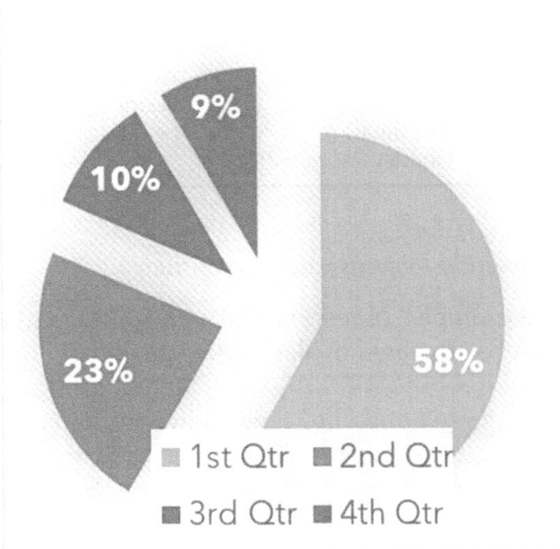

Figure 11.1 Pie chart of quarterly sales by percentage

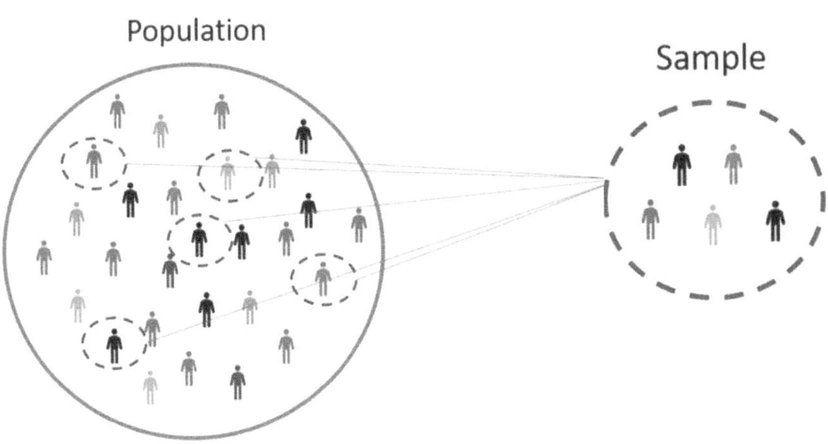

Figure 11.2 Sample is part of the targeted population

Table 11.2 Example of descriptive analysis and data presentation

Research questions/ objective	*Variable name (scale of measurement)*	*Statistics*	*Suggested data presentation*
What is your preferred third-party delivery company? (E.g. DHL, FedEx, GDEX, etc.)	Preferred third-party delivery company (Nominal)	Frequency, mode	Bar graph
How much do you spend on online shopping this month?	Amount spent on online shopping (Interval)	Frequency, mean, mode, standard deviation, variance	n/a
To know the preference of store layout types by gender.	Store layout preference (Nominal) Gender (Nominal)	Cross-tabulation, frequency, percentage	Histogram, box-plot

11.2.1 Research questions/objective/purpose of study

The following are examples of research questions (RQ) and research objectives (RO) that can be answered by **descriptive analysis.**

- RQ1: What is your preferred third-party delivery company?
- RQ2: How much do you spend on online shopping this month?
- RO1: To know the preference of store layout types by gender.

The following can be tested by **hypothesis testing.**

- RO2: To identify the **significant differences** between gender and shopping behaviour.
- RO3: To identify the **significant relationship** between salary and savings.

11.2.2 Questionnaire design/scale of measurement

Questionnaire design

- Single question
- Multiple questions (especially multidimensional variables)

Scale of measurement

- Nominal
- Ordinal
- Interval
- Ratio

Example of questionnaire design

Single question

What is your monthly salary?

Multiple questions

More than one question to assess one variable or the variable has more than one dimension. Thus, the researcher needs to do certain index or score tabulation before data analysis.

E.g. BMI index; quality of life; happiness

11.2.3 Assumption testing

Perform a **normality test** (to assess whether the data are normally distributed) or an **analysis of variance** using a statistical tool such as SPSS. Check the results of the exploratory data analysis.

If your data are normally distributed, the assumption of normality is met, and a parametric test would be appropriate.

Choose parametric tests

E.g. T-tests, correlation

Otherwise, **choose non-parametric tests,** e.g., the Wilcoxon signed-rank test or the Mann-Whitney U test.

11.3 Example of research objectives and types of analysis applied

RO2 and RO3 are examples of research objectives that need hypothesis testing. Assuming the data is normally distributed.

RO2: To identify the significant **difference** between gender and shopping behaviour.

PURPOSE: Test the difference

INDEPENDENT VARIABLE: **Gender** [Male/Female: Nominal scale]
DEPENDENT VARIABLE: **Shopping behaviour** [Total score of multiple questions: Interval]

Refer to Table 11.3, the **independent-samples T-test** is suitable to test the R02.

RO3: To identify the significant **relationship** between salary and savings.

PURPOSE: Test the relationship

INDEPENDENT VARIABLE: Salary [0 – infinity, **ratio**]
DEPENDENT VARIABLE: Savings [0 – infinity, **ratio**]

Based on Table 11.3, **correlation** is suitable to test the R03.

Table 11.3 Summary of the guide for choosing types of data analysis

Purpose (To identify the . . .)	*Requirement of independent variable*	*Requirement of dependent variable*	*Type of analysis*
Parametric test			
Mean differences	Nominal/ordinal (only 2 categories)	Interval/ratio	**Independent-samples T-test**
Mean differences	Nominal/ordinal (> 2 categories)	Interval/ratio	**One-way ANOVA**
Relationship	Interval/ratio	Interval/ratio	**Correlation**
Determinants/ predictors	Interval/ratio/ dummy	Interval/ratio/ dummy	**MLR**
Non-parametric test			
Differences	Ordinal/nominal	Chi-square (non-parametric test)	**Chi-square (test of homogeneity)**
Association/ relationship/ independence	Ordinal/nominal	Chi-square (non-parametric test)	**Chi-square (test of independence)**

By right, you should know which data analysis method to use.
BUT your possible question now is . . .

- What should I do after data entry?
- What can I do with my database?
- Is my data ready to be analysed?

You shall perform exploratory data analysis to get an overview of the database, checking for:

- Descriptive statistics to get an overview of the data set
- Missing data and outliers
- Assumptions testing, such as the normality of the data set

References

Cattell, R. B. (1966). The scree test for the number of factors. *Multivariate Behavioral Research, 1*(2), 245–276.

Child, D. (2006). *The essentials of factor analysis* (3rd ed.). Continuum International Publishing Group, New York, NY.

Cohen, J. (1988). *Statistical power analysis for the behavioral sciences* (2nd ed.). Lawrence Erlbaum Associates.

Field, A. (2024). *Discovering statistics using IBM SPSS statistics*. Sage Publications Limited.

Gabriel, Z., & Bowling, A. (2004). Quality of life from the perspectives of older people. *Aging & Society, 24*, 675–691.

George, D., & Mallery, P. (2024). *IBM SPSS statistics 29 step by step: A simple guide and reference*. Routledge.

Graf, C. (2009). The Lawton instrumental activities of daily living (IADL) scale. *The Gerontologist, 9*(3), 179–186.

Guilford, J. P. (1950). *Fundamental statistics in psychology and education* (2nd ed.). McGraw-Hill.

Joseph Jr, F., William, C., Barry, J. B., & Rolph, E. A. (2009). *Multivariate data analysis* (7th ed.). Pearson Education Limited, Upper Saddle River, NJ.

Kaiser, H. F. (1960). The application of electronic computers to factor analysis. *Educational and Psychological Measurement, 20*(1), 141–151.

Nunnally, J. (1978). *Psychometric Theory* (2nd ed.). New York: McGraw-Hill.

Osborne, J. (2002). Notes on the use of data transformations. *Practical Assessment, Research, and Evaluation, 8*(1).

Tabachnick, B. G., & Fidell, L. S. (2001). *Using multivariate statistics* (4th ed.). Allyn and Bacon, Boston.

Tavakol, M., & Wetzel, A. (2020). Factor analysis: A means for theory and instrument development in support of construct validity. *International Journal of Medical Education, 11*, 245.

Venkatesh, V., Morris, M. G., Davis, G. B., & Davis, F. D. (2003). User acceptance of information technology: Toward a unified view. *MIS Quarterly, 27*(3), 425–478.

Yeoh, S. F., & Chan, Y. F. (2011). Internet banking adoption in Kuala Lumpur: An application of UTAUT model. *International Journal of Business and Management, 6*(4), 161–167.

Yeoh, S. F., Oxley, J., Ibrahim, R., Hamid, T. A., & Syed Abd. Rashid, S. N. (2018). Measurement scale development for mobility-related quality of life among older Malaysian drivers. *Ageing International, 43*(3), 265–278.

Yeoh, S. F., Sharifah, S. A. R., & Haslinda, A. (2009, October). *Driving practices of older Malaysian drivers: The influence of knowledge, attitude and confidence*. National Conference on Postgraduate Research (Ncon-Pgr09), p. 183.

Index

For Product Safety Concerns and Information please contact our EU representative GPSR@taylorandfrancis.com Taylor & Francis Verlag GmbH, Kaufingerstraße 24, 80331 München, Germany